Fracking

# ISSUES IN ENVIRONMENTAL SCIENCE AND TECHNOLOGY

TITLES IN THE SERIES:

*How to obtain future titles on publication:*

A subscription is available for this series. This will bring delivery of each new volume immediately on publication and also provide you with online access to each title via the Internet. For further information visit http://www.rsc.org/issues or write to the address below.

For further information please contact:
Sales and Customer Care, Royal Society of Chemistry, Thomas Graham House, Science Park, Milton Road, Cambridge, CB4 0WF, UK
Telephone: +44 (0)1223 432360, Fax: +44 (0)1223 426017, Email: booksales@rsc.org
Visit our website at www.rsc.org/books

ISSUES IN ENVIRONMENTAL SCIENCE AND TECHNOLOGY

EDITORS: R.E. HESTER AND R.M. HARRISON

39
# Fracking

Issues in Environmental Science and Technology No. 39

Print ISBN: 978-1-84973-920-7
PDF eISBN: 978-1-78262-055-6
ISSN: 1350-7583

A catalogue record for this book is available from the British Library

Published by The Royal Society of Chemistry,
Thomas Graham House, Science Park, Milton Road,
Cambridge CB4 0WF, UK

Registered Charity Number 207890

For further information see our web site at www.rsc.org

Printed and bound by CPI Group (UK) Ltd, Croydon, CR0 4YY

# Preface

The depletion of reserves of conventional natural gas has led to the development of a new technology involving deep horizontal drilling and the injection of high pressure water containing various chemicals and sand in
order to release unconventional gas from underground shale deposits.
Hydraulic fracturing of shales to release the trapped gas has become
known as 'fracking'. The technology has been developed in the USA in recent years where it has been applied on a commercial scale and is cred-
ited with many benefits, including lowered energy costs and increased
national security associated with reduced dependence on imports of oil
and gas. However, this development has not been without controversy.
Contamination of water supplies, causing minor earthquakes and contrib-
uting to climate change are among the accusations leveled at the fracking
industry.

The positive aspects of fracking in the USA have led to its take-up elsewhere in the world, notably in Europe, Australia and China. The negatives,
however, also are vigorously debated and have been judged unacceptable by
some countries. The UK government has given its strong support but there is
a powerful environmental movement opposing the development and a
growing list of countries that have effectively banned fracking, including
France and Germany. The reasons underlying the controversy are explored in
detail in this book, with contributions from all sides of the argument and
with an international and expert authorship.

The book opens with an introduction and overview of the shale gas development written by Peter Hardy of the UK Institution of Gas Engineers and
Managers. The geological requirements and the technology of deep horizontal drilling are described in considerable detail, as are the additives used
in the water: chemicals to modify pH, surface tension and viscosity, for example, and sand to prop open the fissures in the shale rock caused by injection of fluid at high pressure. Wells are drilled vertically through the rock
strata, including those bearing water (the 'water table'), to reach shale deposits at depths of up to several kilometers. Multiple horizontal channels

Issues in Environmental Science and Technology, 39
Fracking
Edited by R.E. Hester and R.M. Harrison
© The Royal Society of Chemistry 2015
Published by the Royal Society of Chemistry, www.rsc.org

that can extend a further 1–1.5 km are then drilled into the shale. Many such wells are usually required in each location, the number of multi-well pads being typically 9 per square mile in the USA. Thus there is considerable above-ground activity associated with fracking, in addition to drilling, with storage of large quantities of both clean and contaminated water and water additives, gas storage, movement of heavy vehicles in the vicinity, *etc*. Particular attention is given in this first chapter to recent developments in the UK and differences from the US experience are discussed.

Wallace Tyner and his colleagues from the Dept. of Agricultural Economics at Purdue University examine the economic impacts of shale oil and gas in the USA in Chapter 2. Using economic modeling, they conclude that the shale boom will have resulted in a 3.5% increase in GDP over the period 2008–2035, coupled with welfare benefits and employment gains. These gains depend, however, on the restriction of gas exports and are accompanied by higher emissions.

Iain Scotchman has written on the important subject of exploration for unconventional hydrocarbons in Chapter 3. As an industry insider working for Statoil, though writing here in a personal capacity, he describes the technical requirements for shale 'plays' to be commercially exploitable. Estimated ultimate recovery factors are given for both biogenic and thermogenic shales and the different structures of the so-called hybrid reservoirs from which shale ('tight') oil may be extracted also are discussed. Areas of potential shale resources are defined by screening of outcrop geochemical data, augmented by test drill cuttings and seismic attributes, with the aim of identifying 'sweet spots' with the characteristics of high organic content, maturity, mineralogy suitable for fracking, thickness and depth.

John Broderick and Ruth Wood of the Tyndall Centre for Climate Change Research at the University of Manchester, UK, consider the climate change impacts of shale gas production in Chapter 4. The internationally agreed target for restricting greenhouse gas emissions to levels needed to avoid dangerous climate change, defined and agreed as more than 2 °C increase in global mean surface temperature above pre-industrial levels, militates against shale gas. There may be a case for its use as a 'transition fuel' but the US experience indicates that even this is unlikely and investment in shale-gas exploration and associated gas infrastructure is, in effect, likely to be an investment in a stranded asset.

The hydrogeological aspects of shale gas extraction in the UK are the subject of Chapter 5 by Robert Ward and colleagues from the British Geological Survey. Given that the areas likely to be exploited for shale gas are overlain in many cases by aquifers used for drinking water supply and for supporting baseflow to rivers, the vulnerability of groundwater and the wider water environment must be taken very seriously. Experience from the US is not encouraging in that many incidences of contamination have been reported. This experience needs to inform a risk management strategy for the UK industry if the potential problems are to be avoided.

In Chapter 6 Alan Randall, Professor of Agricultural and Resource Economics at the University of Sydney, Australia, explores the economic, environmental and policy issues associated with the recovery of gas from coal seams. Although dewatering is the main method used, fracking often is required at least in the later stages of gas extraction and the associated environmental problems are similar to those encountered with shale gas. Empirical findings are presented from a recent case study of the economics of competition and coexistence of coal seam gas and agriculture on prime farmland.

The prospects for shale gas development in China are examined by Shu Jiang of the University of Utah, USA, in Chapter 7. China has the largest estimated quantity of recoverable shale gas in the world and, inspired by the shale gas revolution in the USA, is trying to replicate that country's experience to produce shale gas to power its economy and mitigate the pressure felt from being the world's biggest greenhouse gas emitter. China's government is providing strong financial and policy support for this development but, in addition to the environmental problems encountered elsewhere, difficult geological and water resource conditions need to be faced.

The final chapter is by Tony Bosworth of Friends of the Earth, who claims that shale gas is not only unconventional but also is "unburnable and unwanted". A well-documented case is made for the environmental concerns associated with fracking outweighing any potential benefit and an alternative model for the future of energy supplies is proposed, based on renewables and reduced waste.

In summary, this book presents a critical but balanced and authoritative analysis of developments in the technology, economics, environmental and health concerns associated with fracking. It illuminates the public controversy over the further development and wider application of this process through factual and evidence-based examination of the arguments both in favour of it and against it. The book should be required reading for all concerned with future energy policy development and will be of great value to anyone looking for an in-depth account of this subject which has already attracted so much attention in the popular news media. Students undertaking courses in science, engineering and environmental science will find it of particular value, as will professionals working in the energy industries, the water industry, in land management and climate change.

R. E. Hester
R. M. Harrison

# Contents

Issues in Environmental Science and Technology, 39
Fracking
Edited by R.E. Hester and R.M. Harrison
© The Royal Society of Chemistry 2015
Published by the Royal Society of Chemistry, www.rsc.org

**Exploration for Unconventional Hydrocarbons:**
**Shale Gas and Shale Oil**                                                  **69**
*Iain C. Scotchman*

# Editors

**Ronald E. Hester, BSc, DSc (London), PhD (Cornell), FRSC, CChem**

Ronald E. Hester is now Emeritus Professor of Chemistry in the University of York. He was for short periods a research fellow in Cambridge and an assistant professor at Cornell before being appointed to a lectureship in chemistry in York in 1965. He was a full professor in York from 1983 to 2001. His more than 300 publications are mainly in the area of vibrational spectroscopy, latterly focusing on time-resolved studies of photoreaction intermediates and on biomolecular systems in solution. He is active in environmental chemistry and is a founder member and former chairman of the Environment Group of the Royal Society of Chemistry and editor of 'Industry and the Environment in Perspective' (RSC, 1983) and 'Understanding Our Environment' (RSC, 1986). As a member of the Council of the UK Science and Engineering Research Council and several of its sub-committees, panels and boards, he has been heavily involved in national science policy and administration. He was, from 1991 to 1993, a member of the UK Department of the Environment Advisory Committee on Hazardous Substances and from 1995 to 2000 was a member of the Publications and Information Board of the Royal Society of Chemistry.

**Roy M. Harrison, BSc, PhD, DSc (Birmingham), FRSC, CChem, FRMetS, Hon MFPH, Hon FFOM, Hon MCIEH**

Roy M. Harrison is Queen Elizabeth II Birmingham Centenary Professor of Environmental Health in the University of Birmingham. He was previously Lecturer in Environmental Sciences at the University of Lancaster and Reader and Director of the Institute of Aerosol Science at the University of Essex. His more than 400 publications are mainly in the field of environmental chemistry, although his current work includes studies of human health impacts of atmospheric pollutants as well as research into the chemistry of pollution phenomena. He is a past Chairman of the Environment Group of the Royal Society of Chemistry for whom he has edited 'Pollution: Causes, Effects and Control' (RSC, 1983;

Fourth Edition, 2001) and 'Understanding our Environment: An Introduction to Environmental Chemistry and Pollution' (RSC, Third Edition, 1999). He has a close interest in scientific and policy aspects of air pollution, having been Chairman of the Department of Environment Quality of Urban Air Review Group and the DETR Atmospheric Particles Expert Group. He is currently a member of the DEFRA Air Quality Expert Group, the Department of Health Committee on the Medical Effects of Air Pollutants, and Committee on Toxicity.

# List of Contributors

**John Bloomfield**, British Geological Survey, Maclean Building, Crowmarsh Gifford, Wallingford, Oxfordshire OX10 8BB, UK

**Tony Bosworth**, Friends of the Earth, 26-28 Underwood Street, London N1 7JQ, UK, Email: tony.bosworth@foe.co.uk

**John Broderick**, Tyndall Centre for Climate Change Research, H Floor Pariser Building, University of Manchester, PO Box 88, Manchester M13 9PL, UK, Email: john.broderick@manchester.ac.uk

**Peter Hardy**, Institution of Gas Engineers and Managers (IGEM), IGEM House, 26 & 28 High Street, Kegworth, Derbyshire DE74 2DA, UK, Email: peterangelahardy@gmail.com

**Shu Jiang**, Energy and Geoscience Institute, University of Utah, 423 Wakara Way 300, Salt Lake City UT 84108, USA, Email: sjiang@egi.utah.edu

**Alan Randall**, Agricultural & Resource Economics, Room 218 Biomedical Building, University of Sydney, NSW 2015, Australia, Email: alan.randall@sydney.edu.au

**Kemal Sarica**, Dept of Agricultural Economics, Purdue University, 403 West State Street, West Lafayette IN 47907-2056, USA

**Iain Scotchman**, EXP GNV Unconventional Hydrocarbons, Statoil (UK) Ltd, One Kingdom Street, London W2 6BD, UK, Email: isco@statoil.com

**Marianne Stuart**, British Geological Survey, Maclean Building, Crowmarsh Gifford, Wallingford, Oxfordshire OX10 8BB, UK

**Farzad Taheripour**, Dept of Agricultural Economics, Purdue University, 403 West State Street, West Lafayette IN 47907-2056, USA

**Wallace E. Tyner**, Dept of Agricultural Economics, Purdue University, 403 West State Street, West Lafayette IN 47907-2056, USA, Email: wtyner@purdue.edu

**Robert S. Ward**, British Geological Survey, Environmental Science Centre, Nicker Hill, Keyworth, Nottingham NG12 5GG, UK, Email: rswa@bgs.ac.uk

**Ruth Wood**, Tyndall Centre for Climate Change Research, H Floor Pariser Building, University of Manchester, PO Box 88, Manchester M13 9PL, UK

# Introduction and Overview: the Role of Shale Gas in Securing Our Energy Future

PETER HARDY

## ABSTRACT

The phenomenon of shale gas is both topical and controversial. Its proponents claim that it is a clean, environmentally friendly and abundant source of cheap natural gas; its opponents believe the opposite. In several countries it is a fast-growing industry and operations have begun in the UK.

With conventional reserves of natural gas being quickly depleted, gas prospecting is turning to "unconventional resources", one example being gas found in shale. Uncommon technologies, notably hydraulic fracturing and horizontal drilling, are necessary for shale extraction to be economical.

Shale gas has faced some difficulties over concerns regarding environmental pollution. In the US, *Gasland*, an influential film was released alleging that waste fluid from hydraulic fracturing, "flowback water", was polluting groundwater. While it is possible for methane to enter groundwater through a faulty well completion, in the UK it is the responsibility of the Environment Agency and HSE to ensure regulation is adequate to prevent risks to the environment or human health.

There have been two earthquakes in Lancashire thought to have been caused by shale gas operations. The results of an investigation into these have been accepted as revealing that they were caused by hydraulic fracturing operations and new guidelines are being proposed to reduce the risk of this happening again.

Issues in Environmental Science and Technology, 39
Fracking
Edited by R.E. Hester and R.M. Harrison
© The Royal Society of Chemistry 2015
Published by the Royal Society of Chemistry, www.rsc.org

With insufficient public information and sometimes animosity towards shale gas, drillers need to consider developing corporate social responsibility programs tailored to the needs of the communities local to drilling, with especial consideration towards environmental initiatives.

Worldwide, shale gas has had a significant and growing impact on gas production and looks likely to rapidly transform the energy situation.

In Europe, Poland and France have the largest reserves; Poland has embarked on a program to exploit its shale gas reserves. France, on the other hand, has outlawed the hydraulic fracturing technology vital to shale gas on environmental grounds.

The UK's shale gas reserves are unlikely to be large enough to be a "game changer"; however, they would contribute to gas security and the UK's energy mix, as well as being perceived as a lower-carbon alternative to coal-fired electricity generation.

There are already substantial reserves of gas available worldwide; however, the development of these unconventional gases, which are often in more politically stable parts of the world, may provide a greater security of supply to the Western World going forward.

# 1   Introduction

As the existing conventional gas supplies have started to decline in some parts of the western world, the search has been on for alternative secure sources of supply. One of the most exciting developments in the last 20 years in the natural gas sector has been the development of unconventional gases and, in particular, the exploration and production of shale gas. The existence of shale gas has been known for decades but technological difficulties and substantial financial costs associated with its extraction have up until recently made its exploitation uneconomic. However, increased demand and lagging supply have resulted in the price of gas rising to the point where, along with the development of advances in drilling, shale gas has started to represent a viable alternative to conventional sources of supply. Shale gas is now being produced in large quantities in the USA as their industry has developed over the last decade. Exploitation of reserves is now progressing in other parts of the world, including Poland and Australia. In addition, exploration is starting in other countries including the UK but the development of shale gas production, which often includes hydraulically fracturing of the rock (otherwise known as fracking), is not without its opponents. In America the film, *Gasland*, raised issues relating to problems associated with fracking which has caused some people to have environmental concerns. In some countries, such as France, an embargo has been placed on fracking and even within the USA some states are not as yet permitting it. There have been reports of ground-water contamination which has resulted in illnesses, gas coming out of water taps, and earthquakes caused by fracking. However, in many parts of the world shale gas is seen as a secure source of hydrocarbon that cannot be ignored.

Development is seen in many countries as a way to secure energy supplies that is independent of events in the more volatile parts of the world where most of the existing gas and oil reserves are located.

## 1.1  History

In 1821 shale gas was produced from a natural seepage in the Appalachian Mountains at Fredonia, New York, USA. It was trapped and piped in hollow logs where it was used to light homes and businesses. The profit margins were small and small local operators exploited it as a "cottage industry".[1]

In the late 1960s and early 1970s it was clear that the political situation in the Middle East was changing. There had been Arab–Israeli wars in 1967 and 1973 and the situation led to dramatically increased prices for oil as well as supply shortages. The Organisation of the Petroleum Exporting Countries (OPEC)[†] also rose to international prominence during the 1970s, as its Member Countries took control of their domestic petroleum industries and acquired a major say in the pricing of crude oil on world markets. On two occasions, oil prices rose steeply in a volatile market, triggered by the Arab oil embargo in 1973 and the outbreak of the Iranian Revolution in 1979. OPEC broadened its mandate with the first Summit of Heads of State and Government in Algiers in 1975, which addressed the plight of the poorer nations and called for a new era of cooperation in international relations in the interests of world economic development and stability. This led to the establishment of the OPEC Fund for International Development in 1976. Member Countries embarked on ambitious socio-economic development schemes. It was against this background of volatile oil prices and trying to ensure security of supply that, in 1976, the United States Department of Energy initiated the Eastern Shales Project at a cost of up to $200 million to evaluate the geology, geochemistry and petroleum production engineering of non-conventional petroleum, including shale gas. Important reports established findings from what was then the only shale gas production in the world,[2] based on the Devonian and Mississippian shales in the Appalachian basin.[3] These reports led to the establishment of the Gas Research Institute and also stimulated research at Imperial College in the United Kingdom looking at evaluating potential resources. The geology of the plate tectonics of the Atlantic Ocean implied that the continuation of the Appalachian basin extended across into the UK and on into mainland Europe (see Figure 1). Imperial College focussed on the US paradigm of "cottage industry" and reviewed potential shale gas extraction from throughout the rock strata.[4] The study concluded that Pre-Cambrian and Lower Palaeozoic shales were generally too metamorphosed to be potential reservoirs and most Mesozoic and younger organic-rich shales and mudstones were deemed too immature to

---

[†]OPEC is a permanent intergovernmental organisation, created at the Baghdad Conference on September 10–14, 1960, by Iran, Iraq, Kuwait, Saudi Arabia and Venezuela; membership grew to 13 by 1975.

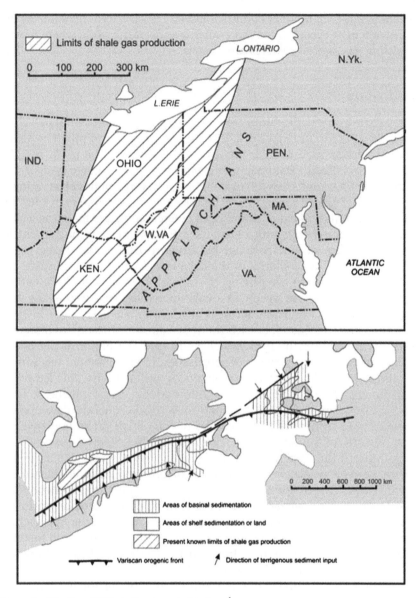

**Figure 1** Limits of US shale gas production.[1]

be considered. Carboniferous shales, in general, and Namurian shales, in particular, were found to be ideally suited, both in terms of maturity and in degree of natural fracturing (see Figure 2). During the late 1970s, profit made from gas extraction was subject to both Corporation Tax and Petroleum Revenue Tax, meaning that production was nowhere near economic.

The conclusions of the Imperial College study on shale gas potential in the UK were presented to the UK Department of Energy in 1985. They were met

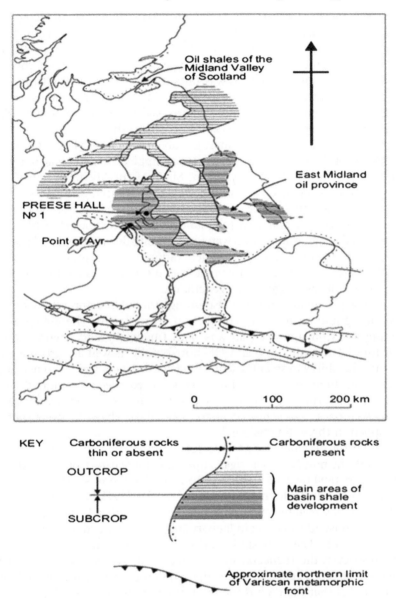

**Figure 2**   Location of shale-gas-bearing strata in the United Kingdom.[1]

with polite interest but the chances of shale gas being exempt from Petroleum Revenue tax was not countenanced. Subsequent attempts to inform the wider world failed and no reputable scientific journal would publish papers on the UK's shale gas resources. Finally, conclusions of the research were published in the US.[4] Meanwhile in the US, shale gas development was continuing in the Appalachians, from a geographic perspective as well as from the study of rock strata, especially the distribution, deposition and age of sedimentary rocks, and various technological advances were also being looked at. The Appalachian basin, from New York State through Ohio to Kentucky and Illinois, was the main historic area for shale gas production, but there had been other basins where the gas was produced, such as the Williston Basin. This is a large intracratonic sedimentary basin in eastern Montana, western North Dakota, South Dakota and southern Saskatchewan where the Bakken Shale had produced gas since 1953. Stimulated by the Department of Energy and the Gas Research Institute, shale gas areas were found in the Cretaceous Lewis Shale of the San Juan Basin, the Mississippian Barnett Shale of the Fort Worth Basin and the Devonian Antrim Shale of the Michigan Basin.[5] Geochemical studies revealed that the gas was not produced by microorganisms that generate heat within organic waste (thermogenic) but was produced by bacterial anaerobic respiration (methanogenesis). The bacteria had entered the fractured shale from ground water percolating down from the glacial drift cover.[6] This second process for gas generation opened up new areas for exploration: areas where the source rock was previously deemed immature or over-mature for thermogenic gas generation.

The shale gas renaissance was also brought about by improved methods of well drilling and advances in completion technologies. The ability to drill multiple wells off a single pad was both financially and environmentally rewarding.[7] The ability to drill wells horizontally as well as vertically, together with the ability to steer the drill along "sweet spots", enabled permeable gas-changed zones to be tapped into. This was coupled with more dramatic hydraulic fracturing techniques. Seismic techniques, which could use the fracturing process as an energy source, enabled gas-charged "sweet spots" to be mapped in three dimensions.[1]

In the US the development of shale gas expanded dramatically from the mid 1990s, with the number of gas rigs in operation increasing from around 250 in 1993 to over 1500 by 2007.[8] This has seen production of shale gas in the US increase from 1293 billion cubic feet in 2007 to 7994 billion cubic feet in 2011.[9] This has resulted in the US natural gas wellhead price falling from $8.01 per thousand cubic feet in January 2006 to $2.89 per thousand cubic feet in January 2012.[9] These are similar prices to those seen in the US in the early 1980s. However, the reduction in the price of natural gas seen in the US is unlikely to be repeated to the same extent in Europe. This is due to the limited export market that the US has for gas at the present time as it has no export terminals for shipping the gas globally as liquefied natural gas (LNG). The largest exporter of LNG worldwide, Qatar, has six operational export terminals, whilst Australia, which is rapidly increasing its export of LNG, has three operational export terminals with another seventeen projects either under construction or being planned. Worldwide there are thirty-two operational export terminals with another sixty-nine under construction or in planning.[10]

In the UK, shale gas as a potential industry did not develop at all until the British Geological Survey (BGS) noted the potential for its production in 1995.[11] Shale gas was not mentioned in reviews of future UK petroleum resources published in 2003 by the Oil and Gas Directorate of the Department of Trade & Industry.[12] The 6th *Petroleum Geology Conference on the Global Perspectives of North West Europe* was held later in the same year. The three-day programme concluded with a session on non-conventional petroleum. This included a presentation on the UK's shale gas resources and provided a platform to disseminate updated conclusions of the Imperial College research of some 15 years previously. The advances in US shale gas exploration and production technology could now be applied to the UK.[13] In 2008 the British Geological Survey began to review UK shale gas resources and delivered a presentation on their results at the 7th *Petroleum Conference* in March 2009. Subsequently, the Department of Energy & Climate Change commissioned the BGS to prepare a report on *The Unconventional Hydrocarbon Resources of Britain's Onshore Basins – Shale Gas.*[14]

The result of this was that several companies started to look at shale gas sites within the UK at the time of the announcement of the 13th onshore round of UK licences in 2006. In 2008 Wealden Petroleum Developments Ltd was awarded a license that covered large parts of the Weald, an area in South East England situated between the parallel chalk escarpments of the North and the South Downs, for exploration. Additionally, Cuadrilla Resources Corporation was awarded a licence that includes areas of the North West of England.

## 2 Shale Gas Production and Reserves in the UK

### 2.1 Overview

*2.1.1 Shale Gas Production Process.* As noted, horizontal drilling and hydraulic fracturing are the two technologies that together have the potential to unlock the tighter shale gas formations.

Hydraulic fracturing (also known as "fracking") is a well-stimulation technique which consists of pumping a fluid and a propping agent ("proppant"), such as sand, down the wellbore under pressure to create fissures in the hydrocarbon-bearing rock. Propping agents are required to "prop open" the fracture once the pumps are shut down and the fracture begins to close. The ideal propping agent is strong, resistant to crushing, resistant to corrosion, has a low density and is readily available at low cost. The products that best meet these desired traits are silica sand, resin-coated sand (RCS) and ceramic proppants. The fractures start in the horizontal wellbore and can extend for several hundred metres while the sand holds the fissures apart, allowing the gas to flow into the wellbore. Recovery of the injected fluids is highly variable, depending on the geology, and ranges from 15 to 80%.[15]

Horizontal drilling allows the well to penetrate into the hydrocarbon rock seam which can be typically 90 m thick in the US, but can be up to 1000 m thick in some of the UK shale gas seams. Horizontal drilling maximises the rock area that, once fractured, is in contact with the wellbore and therefore maximises the volume of shale gas that is released.

Horizontal drilling is performed with similar equipment and technologies to that which has been established over decades for vertical drilling and, in fact, the initial drilling of the vertical bore is almost identical to a conventional well. However, the well development and gas extraction processes differ widely between conventional and unconventional gas production. Whilst some conventional wells have been stimulated by hydraulic fracturing in the past, horizontal drilling and hydraulic fracturing are key requirements to make the exploitation of shale gas deposits economically viable.

The requirement for horizontal drilling and hydraulic fracturing also results in differences in the distribution of wells above the shale gas formations. The process involves locating several individual wells on a single "multi-well" pad. Normally 6–10 horizontal wells radiate out from the centre well pad; these then are drilled in parallel rows, typically 5–8 m apart. Each horizontal wellbore is typically 1–1.5 km in lateral length, although they can be considerably longer.

As the array of wells drilled from each pad only enables access to a discrete area of the shale formation, several multi-well pads in a geographic area are required in order to maximise shale gas extraction. In the US they typically locate a maximum of nine pads per square mile. In the UK, Composite Energy has estimated that about three pads per square mile should be sufficient for the UK setting.[16] However, the geological and above-ground constraints will also impact on the location of well pads.

The differences in the production process between conventional and unconventional gas production also results in differences in the level of effort required to extract shale gas. It also affects the amount of resources used and the corresponding volume of waste products generated.

*2.1.2  Well Pad Construction.*  As already stated previously, the pads used for multi-well drilling require an area of land sufficient to accommodate fluid storage and equipment associated with the hydraulic fracturing operation. This utilises larger equipment for horizontal drilling than that required for vertical drilling only. This results in between 0.4 and 1.2 ha (1–3 acres) of land being required for a multi-well pad.

*2.1.3  Drilling.*  Vertical drilling depth will vary dramatically, depending on the depth of the shale gas strata and their location. However, it is expected that wells will be drilled through rock layers and aquifers for a distance of up to 2 km, to within 150 m of the top of the shale gas rock to be hydraulically fractured. A more powerful horizontal drill may then be used for the horizontal portion of the wellbore. This transition is known as "kicking off" and the horizontal well is then continued for an additional 1–1.5 km.

The vertical portion of the well is typically drilled using either compressed air or freshwater mud as a drilling fluid. Once the horizontal section is ready to be drilled, then this normally requires drilling mud for powering and cooling the down-hole motor that is used for the directional drilling. The drilling mud also provides stability for the horizontal drilling and for using the navigational tools, which require mud to transmit the sensor readings that enable the bore to be accurately traced. The mud also enables the removal of the cut material from the drilling operation. The drilling mud is a heavier mud than the freshwater mud used in the vertical drilling due to the need to prevent hole collapse of the horizontal wellbore where the earth's vertical stress is much greater than in the vertical bore.

Developments are being undertaken to drill the horizontal bore pneumatically, using specialist equipment to control fluids and gases entering the wellbore. It will be interesting to see which method will come to dominate the horizontal drilling operation, although it could be the geology or water availability that determines the approach adopted.

In terms of waste material generated from the process, if the vertical well was 2 km deep with a 1.2 km lateral well the volume of waste would be in the region of 140 cubic metres, whereas a conventional well drilled to the same depth of 2 km would generate about 85 $m^3$ of waste material. Therefore a 10-well pad would generate 1400 $m^3$ of waste.

*2.1.4  Well Casings.*  Well casings[17] are installed to seal the well from the surrounding formation and to stabilise the completed well. A number of these may be installed to meet a variety of circumstances and are typically concentric steel pipes lining the inside of the drilled hole, with the annular spaces filled with cement. There are four casing "strings", each installed at different stages in drilling. The first is the "conductor casing" this is installed during the first phase of drilling; it is a shallow steel conductor casing installed vertically to reinforce and stabilise the ground surface, the depth of the conductor casing is typically 40–300 ft. This is followed by the "surface casing". After the installation of the conductor casing, drilling continues to below the bottom of the freshwater aquifers (depth requirements for groundwater protection are likely to be the subject of approval from the Environment Agency in the UK), at which point a second casing of smaller diameter (the surface casing) is installed and cemented in (see Figure 3).

Intermediate casings of still smaller diameter are sometimes installed from the bottom of the surface casing to a deeper depth. This is usually only required for specific reasons such as additional control of fluid flow and pressure effects or for protection of other resources such as minable coals or gas storage zones. It could, of course, form part of the requirements from the regulatory authorities in the UK.

Regulation in the UK is controlled by a variety of national and local government departments with a variety of different responsibilities. In the UK the environmental considerations are controlled by the following organisations: The Environmental Agency (EA) in England, the Scottish

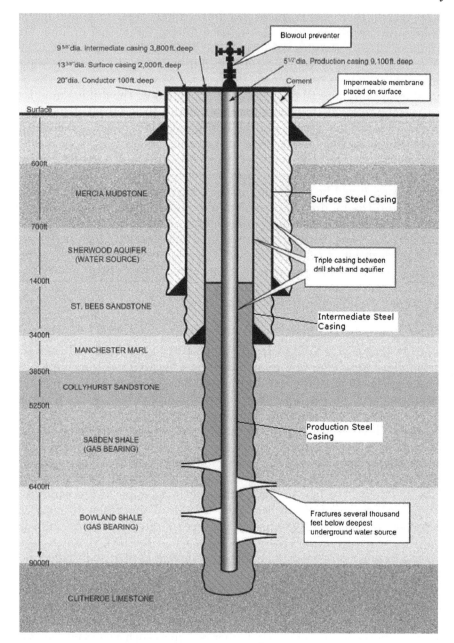

**Figure 3**   Cross section of typical well casings.[17]

Environmental Protection Agency (SEPA) in Scotland and the Environmental Agencies of Wales and Northern Ireland. The Department of Energy and Climate Change administers the licensing system while the Planning

Authorities (generally the local authority) deal with the planning applications required for each site.

After the surface casing cement is set (and intermediate casings, if required) the well is drilled to the target shale gas formation and a "production casing" is installed, either at the top of the target formation or into it, depending on the type of well being installed (either "open hole" or "through-perforated casing", respectively). Well completions incorporate the steps taken to transform a drilled well into a producing one. These steps include casing, cementing, perforating and installing a production tree.

Open hole completions are the most basic type and are used in formations that are unlikely to cave in. An open hole completion consists of simply running the casing directly down into the formation, leaving the end of the piping open without any other protective filters. Very often, this type of completion is used on formations that have been fractured.

Conventional through-perforated completions consist of production casing being run through the formation. The sides of this casing are perforated, with small holes along the sides facing the formation, which allows for the flow of hydrocarbons into the well hole but still provides a suitable amount of support and protection for the well hole. The process of perforating the casing involves the use of specialised equipment designed to make small holes through the casing, cementing, and any other barrier between the formation and the open well. In the past, "bullet perforators" were used, which were essentially small guns lowered into the well. The guns, when fired from the surface, sent off small bullets that penetrated the casing and cement. Today, "jet perforating" is preferred. This consists of small, electrically ignited charges that are lowered into the well. When ignited, these charges blast small holes through to the formation, in the same manner as bullet perforating.

In addition to the depth of the surface casing, the regulatory authorities in the UK are likely to put requirements on the cementing-in of the surface casing. A method known as "circulation" may be used to fill the entire space between the casing and the wellbore (the annulus or outer space between the well casing and the rock through which it has been drilled) from the bottom of the surface casing to the ground surface. Here cement is pumped down the inside of the casing forcing it up from the bottom of the casing into the space between the outside of the casing and the wellbore. Once a sufficient volume of cement to fill the annulus has been pumped into the casing, it is usually followed by pumping a volume of water into the casing to push the cement back up the annular space until the cement begins to appear at the surface. Once the cement appears at the surface the pumping of water is stopped, this ensures that the top section of the annular space is fully filled with cement and therefore there is no leakage path between the outside of the well casing and the rock through which it has been drilled. This method is regarded as the highest standard of cementation compared to other methods such as cementing of the annular space across only the deepest groundwater zone but not all groundwater zones.

Once the surface casing is in place, the regulatory authorities may require operators to install blowout-prevention equipment at the surface, to prevent any pressurised fluids encountered during drilling from moving through the space between the drill pipe and the surface casing.

The operators could also be required to completely fill the annulus with cement from the bottom to the top of the production casing. However, there could be reasons why full cementation is not always required, including the fact that in very deep wells the "circulation" technique of filling the annular space with cement (see previously) is more difficult to accomplish as ce-menting must be handled in multiple stages, which can result in a poor cement job or damage to the casing.

In some instances, well tubings are inserted inside the above casings. They are typically of steel pipe, but they are not usually cemented into the well.

Prior to fracturing, the last step is to install the wellhead that is designed and pressure-rated for the specific hydraulic fracturing operation. In add-ition to proving the equipment for pumping and controlling fluid pressure, the wellhead incorporates flowback equipment to deal with the flowback of fracturing fluid from the well and includes pipes and manifolds connected to a gas–water separator and tanks. Figure 4 shows a typical well site during a single hydraulic fracturing operation.

*2.1.5 Fracturing Procedure.* The fracturing procedure is carried out one well at a time, with each well having multiple stages. A multi-stage proced-ure begins with isolating the well to be fractured and fracturing portions of the horizontal bore, starting at the far end of the wellbore by pumping fracturing fluid in and maintaining high pressures. The process is then re-peated for the next section back and typically the process to hydraulically fracture a 1.2 km horizontal bore consists of 8–13 fracture stages. The pressures used in the fracturing process range from 5000–10000 psi. To minimise the risk of any issues when using the hydraulic fracturing fluid, the operator can pump water and mud into the bore to test the production casing to at least the maximum anticipated treatment pressure. Test pres-sures above the maximum treatment pressures can be used but they should never exceed the casing's internal yield pressure.

The whole of the above process from construction of the access road to the well clean-up and testing can take between 500 and 1500 days, based on the experience in New York State as shown in Table 1.[15]

## 2.2 Exploitation of Reserves

*2.2.1 Ownership of Reserves.* In the USA the onshore shale gas industry has developed with very little in the way of public dissent. The reasons for this include the fact that the land owner often owns the mineral rights as well. This has resulted in many of the landowners actively encouraging explor-ation and production as they can ensure contractual arrangements for a

| | |
|---|---|
| 1. Well head and frac tree with 'Goat Head' | 11. Frac additive trucks |
| 2. Flow line (for flowback & testing) | 12. Blender |
| 3. Sand separator for flowback | 13. Frac control and monitoring center |
| 4. Flowback tanks | 14. Fresh water impoundment |
| 5. Line heaters | 15. Fresh water supply pipeline |
| 6. Flare stack | 16. Extra tanks |
| 7. Pump trucks | 17. Line heaters |
| 8. Sand hogs | 18. Separator-meter skid |
| 9. Sand trucks | 19. Production manifold |
| 10. Acid trucks | |

**Figure 4**   A well site during a single hydraulic fracturing operation.[15]

percentage of the value of the gas produced. This has resulted in the landowners acquiring considerable personal wealth from the development of shale gas wells. The population density in the areas of production is often very low, so few people are affected by the issues associated with the development of the exploration and production wells.

In the UK the situation is entirely different. The mineral rights in most of the UK are owned by the state and, as such, while landowners can get money for the disruption and loss of crops, the value the landowners receive is small compared with the US. A large proportion of areas where shale gas is likely to occur are close to centres of population and therefore the potential impact on the community is greater. The initial development of the sites often involves considerable heavy vehicle movements which impacts on the local users of roads that are rural in nature and which were not designed for the number and size of the vehicles involved. The resulting noise and dust is likely to affect local residents as the exploratory well is constructed.

**Table 1** Summary of mechanical operations prior to production.[15]

| Operation | Materials and equipment | Activities | Duration |
|---|---|---|---|
| Access road and well-pad construction | Backhoes, bulldozers and other types of earth-moving equipment. | Clearing, grading, pit construction, placement of road materials such as geotextile and gravel. | Up to 4 weeks per well pad |
| Vertical drilling with smaller rig | Drilling rig, fuel tank, pipe racks, well control equipment, personnel vehicles, associated outbuildings, delivery trucks. | Drilling, running and cementing surface casing, truck trips for delivery of equipment and cement. Delivery of equipment for horizontal drilling may commence during late stages of vertical drilling. | Up to 2 weeks per well; one-to-two wells at a time |
| Preparation for horizontal drilling with larger rig | | Transport, assembly and setup, or repositioning on site of large rig and ancillary equipment. | 5–30 days per well |
| Horizontal drilling | Drilling rig, mud system (pumps, tanks, solids control, gas separator), fuel tank, well control equipment, personnel vehicles, associated outbuildings, delivery trucks. | Drilling, running and cementing production casing, truck trips for delivery of equipment and cement. Deliveries associated with hydraulic fracturing may commence during late stages of horizontal drilling. | Up to 2 weeks per well; one-to-two wells at a time |
| Preparation for hydraulic fracturing | | Rig down and removal or repositioning of drilling equipment. Truck trips for delivery of temporary tanks, water, sand, additives and other fracturing equipment. Deliveries may commence during late stages of horizontal drilling. | 30–60 days per well, or per well pad if all wells treated during one mobilisation |

| Hydraulic fracturing procedure | Temporary water tanks, generators, pumps, sand trucks, additive delivery trucks and containers, blending unit, personnel vehicles, associated outbuildings, including computerised monitoring equipment. | Fluid pumping, and use of wireline equipment between pumping stages to raise and lower tools used for downhole well preparation and measurements.[a] Computerised monitoring. Continued water and additive delivery. | 2–5 days per well, including approximately 40–100 hours of actual pumping |
|---|---|---|---|
| Fluid return (flowback) and treatment | Gas/water separator, flare stack, temporary water tanks, mobile water treatment units, trucks for fluid removal if necessary, personnel vehicles. | Rig down and removal or repositioning of fracturing equipment; controlled fluid flow into treating equipment, tanks, lined pits, impoundments or pipelines; truck trips to remove fluid if not stored on site or removed by pipeline. | 2–8 weeks per well, may occur concurrently for several wells |
| Waste disposal | Earth-moving equipment, pump trucks, waste transport trucks. | Pumping and excavation to empty/reclaim reserve pit(s). Truck trips to transfer waste to disposal facility. | Up to 5 weeks per well pad |
| Well clean-up and testing | Well head, flare stack, waste water tanks. Earthmoving equipment. | Well flaring and monitoring. Truck trips to empty waste water tanks. Gathering line construction may commence if not done in advance. | 0.5–30 days per well |

**Overall duration of activities for all operations (prior to production) for a six-well multi-well pad**  500–1500 days

[a]Wireline equipment refers to cabling technology used by operators to lower equipment or measurement devices into the well for the purposes of well intervention, reservoir evaluation and pipe recovery.

*2.2.2 Environmental Issues.* There are a number of environmental concerns, both real and imaginary, relating to the exploitation of shale gas in the UK, which, due to the fact that we live on a relatively small island which is heavily populated, means that the development of shale gas is invariably going to impact one community or another.

This could be in the exploration and production phase of the work or in the need to lay gas pipelines to access the existing gas distribution network. It is these issues and how the shale gas industry and the Government address them that will play a key role in allaying public concerns and convincing the public that the resources can be developed in a safe and efficient manner.

*2.2.2.1 Hydraulic Fracturing can have Adverse Effects on Drinking Water*
Vertical drilling is a well-established practice that has been carried out over many years and millions of wells have been drilled through aquifers with no significant issues. Drinking water aquifers are normally at depths of 300 m or less while the natural-gas-producing shale formations are typically at 3000–4000 m. Wells have metal casings between the rock and the bore, which extend well below the levels of the aquifers, and the gaps between the rock and the casings are filled with cement. The design of the casings is required to take account of the geology of the site and any fluids within them and, if necessary, there can be multiple casings extending below the drinking water aquifers to reduce the possibility of contamination. In the extremely rare cases where groundwater has been contaminated it was found to be as a result of faulty well casing installations.

There are a number of precautions that can be taken to minimise the risk to groundwater, in addition to the design and construction of the well. These include monitoring the water quality before and during the operation, having a quality assurance programme to ensure that the equipment and materials are to the correct specification and maintaining close supervision while the work is carried out. A minimum well depth can also be set to ensure adequate separation of the aquifer and the shale to be hydraulically fractured.

*2.2.2.2 Water Volumes used for Hydraulic Fracturing*
Concerns have been raised at the large quantities of water that are used in the process of hydraulic fracturing, particularly in areas such as the South East of England where the existing water infrastructure is under stress. The volumes required for hydraulically fracturing a single well are in the region of 10–20 million litres of water, depending on well depth, length and geology. For a typical drill pad consisting of 10 wells, this will require 100–200 million litres of water per pad. In reality, compared to the daily usage of water in the UK of 15 000 million litres per day,[18] this represents a relatively small volume. Sourcing and use of water is heavily regulated and therefore the amount that can be abstracted at any time will be closely monitored and

controlled to ensure that it does not have an adverse effect on other users. It should be noted that in some of the areas where shale gas deposits are located, such as the North West of England, there is abundant water and therefore the quantities of water used will have little impact on the overall water supply situation.

The cost of water, as well as its availability, will ensure that the shale gas industry is constantly attempting to reduce the volumes used by improving the hydraulic fracturing process, as well as re-using the water wherever possible to mitigate overall water requirements.

### 2.2.2.3  Hydraulic Fracturing Fluids

While the majority of the fluids used for hydraulic fracturing consist of more than 99.5% water and sand, companies do use a small quantity of additional chemicals to assist in the process. Although the percentage of chemical additives is small, this still equates to some 1000–3000 tonnes of chemicals for hydraulically fracturing a typical eight-well pad. The water and additives are blended on site and, when mixed with the proppant, usually sand, are pumped into the wellbore. Chemicals perform many functions in a hydraulic fracturing job. Although there are dozens to hundreds of chemicals which could be used as additives, there are a limited number which are routinely used in hydraulic fracturing. Table 2 shows a list of the chemicals used most often in America.[19]

One of the problems associated with identifying chemicals is that some chemicals have multiple names. For example, ethylene glycol (antifreeze) is also known by the names ethylene alcohol, glycol, glycol alcohol, Lutrol 9, Macrogol 400 BPC, monoethylene glycol, Ramp, Tescol, 1,2-dihydroxy-ethane, 2-dydroxyethanol, $HOCH_2CH_2OH$, dihydroxyethane, ethanediol, ethylene gycol, Glygen, Athylenglykol, ethane-1,2-diol, Fridex, MEG, 1,2-ethandiol, Ucar 17, Dowtherm SR 1, Norkool, Zerex, aliphatic diol, Ilexan E, ethane-1,2-diol and 1,2-ethanediol.

These additives are there for a number of reasons, such as helping dissolve minerals and initiate fissures; preventing scale deposits in the pipes; eliminating bacteria in the water; minimising friction between the fluid and the pipe; preventing precipitation of metal oxides; and thickening the water to suspend the sand which is used to hold the fissures generated apart. While many of the chemicals used are found in common household and commercial products, such as table salt, food additives and cosmetics, some, used in small quantities, are toxic. The number of additives used varies between 3 and 12 as the composition of the fracturing fluid is individually designed for the shale formation being fractured. In the USA the composition of the hydraulic fracturing fluid has not always been disclosed, with some of the companies maintaining that this is commercially sensitive information. This has led to suspicion by members of the public, particularly where health issues have occurred close to shale gas extraction sites. However, in the UK the composition of the fracturing fluid together with all of the

**Table 2** A list of the chemicals used most often in America.[20]

| Chemical name | CAS | Chemical purpose | Product function |
|---|---|---|---|
| Hydrochloric Acid | 007647-01-0 | Helps dissolve minerals and initiate cracks in the rock | Acid |
| Glutaraldehyde | 000111-30-8 | Eliminates bacteria in the water that produces corrosive by-products | Biocide |
| Quaternary Ammonium Chloride | 012125-02-9 | Eliminates bacteria in the water that produces corrosive by-products | Biocide |
| Quaternary Ammonium Chloride | 061789-71-1 | Eliminates bacteria in the water that produces corrosive by-products | Biocide |
| Tetrakis Hydroxymethyl-Phosphonium Sulfate | 055566-30-8 | Eliminates bacteria in the water that produces corrosive by-products | Biocide |
| Ammonium Persulfate | 007727-54-0 | Allows a delayed break down of the gel | Breaker |
| Sodium Chloride | 007647-14-5 | Product Stabiliser | Breaker |
| Magnesium Peroxide | 014452-57-4 | Allows a delayed break down the gel | Breaker |
| Magnesium Oxide | 001309-48-4 | Allows a delayed break down the gel | Breaker |
| Calcium Chloride | 010043-52-4 | Product Stabiliser | Breaker |
| Choline Chloride | 000067-48-1 | Prevents clays from swelling or shifting | Clay Stabiliser |
| Tetramethyl ammonium chloride | 000075-57-0 | Prevents clays from swelling or shifting | Clay Stabiliser |
| Sodium Chloride | 007647-14-5 | Prevents clays from swelling or shifting | Clay Stabiliser |
| Isopropanol | 000067-63-0 | Product stabiliser and/or winterising agent | Corrosion Inhibitor |
| Methanol | 000067-56-1 | Product stabiliser and/or winterising agent | Corrosion Inhibitor |
| Formic Acid | 000064-18-6 | Prevents the corrosion of the pipe | Corrosion Inhibitor |
| Acetaldehyde | 000075-07-0 | Prevents the corrosion of the pipe | Corrosion Inhibitor |
| Petroleum Distillate | 064741-85-1 | Carrier fluid for borate or zirconate crosslinker | Crosslinker |
| Hydrotreated Light Petroleum Distillate | 064742-47-8 | Carrier fluid for borate or zirconate crosslinker | Crosslinker |
| Potassium Metaborate | 013709-94-9 | Maintains fluid viscosity as temperature increases | Crosslinker |
| Triethanolamine Zirconate | 101033-44-7 | Maintains fluid viscosity as temperature increases | Crosslinker |
| Sodium Tetraborate | 001303-96-4 | Maintains fluid viscosity as temperature increases | Crosslinker |
| Boric Acid | 001333-73-9 | Maintains fluid viscosity as temperature increases | Crosslinker |
| Zirconium Complex | 113184-20-6 | Maintains fluid viscosity as temperature increases | Crosslinker |

| Chemical Name | CAS Number | Purpose | Type |
| --- | --- | --- | --- |
| Borate Salts | N/A | Maintains fluid viscosity as temperature increases | Crosslinker |
| Ethylene Glycol | 000107-21-1 | Product stabiliser and/or winterising agent | Crosslinker |
| Methanol | 000067-56-1 | Product stabiliser and/or winterising agent | Crosslinker |
| Polyacrylamide | 009003-05-8 | "Slicks" the water to minimise friction | Friction Reducer |
| Petroleum Distillate | 064741-85-1 | Carrier fluid for polyacrylamide friction reducer | Friction Reducer |
| Hydrotreated Light Petroleum Distillate | 064742-47-8 | Carrier fluid for polyacrylamide friction reducer | Friction Reducer |
| Methanol | 000067-56-1 | Product stabiliser and/or winterising agent | Friction Reducer |
| Ethylene Glycol | 000107-21-1 | Product stabiliser and/or winterising agent | Friction Reducer |
| Guar Gum | 009000-30-0 | Thickens the water in order to suspend the sand | Gelling Agent |
| Petroleum Distillate | 064741-85-1 | Carrier fluid for guar gum in liquid gels | Gelling Agent |
| Hydrotreated Light Petroleum Distillate | 064742-47-8 | Carrier fluid for guar gum in liquid gels | Gelling Agent |
| Methanol | 000067-56-1 | Product stabiliser and/or winterising agent | Gelling Agent |
| Polysaccharide Blend | 068130-15-4 | Thickens the water in order to suspend the sand | Gelling Agent |
| Ethylene Glycol | 000107-21-1 | Product stabiliser and/or winterising agent | Gelling Agent |
| Citric Acid | 000077-92-9 | Prevents precipitation of metal oxides | Iron Control |
| Acetic Acid | 000064-19-7 | Prevents precipitation of metal oxides | Iron Control |
| Thioglycolic Acid | 000068-11-1 | Prevents precipitation of metal oxides | Iron Control |
| Sodium Erythorbate | 006381-77-7 | Prevents precipitation of metal oxides | Iron Control |
| Lauryl Sulfate | 000151-21-3 | Used to prevent the formation of emulsions in the fracture fluid | Non-Emulsifier |
| Isopropanol | 000067-63-0 | Product stabiliser and/or winterising agent | Non-Emulsifier |
| Ethylene Glycol | 000107-21-1 | Product stabiliser and/or winterising agent | Non-Emulsifier |
| Sodium Hydroxide | 001310-73-2 | Adjusts the pH of fluid to maintains the effectiveness of other components, such as crosslinkers | pH Adjusting Agent |
| Potassium Hydroxide | 001310-58-3 | Adjusts the pH of fluid to maintains the effectiveness of other components, such as crosslinkers | pH Adjusting Agent |
| Acetic Acid | 000064-19-7 | Adjusts the pH of fluid to maintains the effectiveness of other components, such as crosslinkers | pH Adjusting Agent |
| Sodium Carbonate | 000497-19-8 | Adjusts the pH of fluid to maintains the effectiveness of other components, such as crosslinkers | pH Adjusting Agent |

**Table 2** Continued.

| Chemical name | CAS | Chemical purpose | Product function |
|---|---|---|---|
| Potassium Carbonate | 000584-08-7 | Adjusts the pH of fluid to maintains the effectiveness of other components, such as crosslinkers | pH Adjusting Agent |
| Copolymer of Acrylamide and Sodium Acrylate | 025987-30-8 | Prevents scale deposits in the pipe | Scale Inhibitor |
| Sodium Polycarboxylate | N/A | Prevents scale deposits in the pipe | Scale Inhibitor |
| Phosphonic Acid Salt | N/A | Prevents scale deposits in the pipe | Scale Inhibitor |
| Lauryl Sulfate | 000151-21-3 | Used to increase the viscosity of the fracture fluid | Surfactant |
| Ethanol | 000064-17-5 | Product stabiliser and/or winterising agent | Surfactant |
| Naphthalene | 000091-20-3 | Carrier fluid for the active surfactant ingredients | Surfactant |
| Methanol | 000067-56-1 | Product stabiliser and / or winterising agent. | Surfactant |
| Isopropyl Alcohol | 000067-63-0 | Product stabiliser and/or winterising agent | Surfactant |
| 2-Butoxyethanol | 000111-76-2 | Product stabiliser | Surfactant |

additives will be fully disclosed. As a result of this and the likely public concern relating to some of the additives, companies working in the UK are likely to invest in "green" or non-toxic alternatives wherever possible.

The fracturing fluid that Cuadrilla has used at the Preese Hall exploration well site, and plans to use at future exploration well sites, is composed almost entirely of fresh water and sand. Cuadrilla also has approval to use the following additives:

- Polyacrylamide (friction reducer)
- Sodium salt (for tracing fracturing fluid)
- Hydrochloric acid (diluted with water)
- Glutaraldehyde biocide (used to cleanse water and remove bacteria)

So far, as additives to fracturing fluid, Cuadrilla has only used poly-acrylamide friction reducer along with a miniscule amount of salt, which acts as a tracer. Cuadrilla have not needed to use biocide as the water supplied by United Utilities to their Lancashire exploration well sites has already been treated to remove bacteria, nor have they used diluted hydrochloric acid in fracturing fluid. Additives proposed, in the quantities proposed, have resulted in the fracturing fluid being classified as non-hazardous by the Environment Agency.

### 2.2.2.4   Hydraulic Fracturing

The issue of hydraulic fracturing causing earthquakes came to prominence in the UK in 2011 when two tremors, one of magnitude 2.3, hit the Fylde coast in Lancashire on the 1st April followed by a second of magnitude 1.4 on the 27th May.

Following investigation by the British Geographical Survey, the epicentre for each earthquake was identified as about 500 metres away from the Preese Hall 1 well at Weeton, Blackpool following hydraulic fracturing.

The geo-mechanical study of the Bowland Shale Seismicity report carried out by independent experts said that the combination of geological factors that caused the quakes was rare and would be unlikely to occur together again at future well sites.[21]

There was no damage as a result of the two earthquakes and the report stated "If these factors were to combine again in the future, local geology limits seismic activity events to around a magnitude 3 on the Richter scale as a worst-case scenario". To put this into context, an earthquake of magnitude 2.5 or less is usually not felt but can be recorded by a seismograph, while earthquakes of magnitude 2.5 to 5.4 are often felt but only cause minor damage. Dr Cliff Frohlich of the University of Texas at Austin carried out a study of the correlation between injection wells and small earthquakes and he commented that there is a question of what kind of damage that a magnitude 3 earthquake could do to drilling infrastructure. "It's plausible that the tremors could affect well integrity," Frohlich says. "In my business, you never say never. That said, most of the time these earthquakes are not right near the well. But it's possible an earthquake could hurt a well," though he knows of no instances where that has occurred.[22]

In order to minimise the risk of future seismic events the companies now review local geology for potential fault lines prior to drilling. In addition, they monitor the process with very sensitive instruments so that the operation can be halted if there are indications that an earthquake is likely to be triggered.

### 2.2.2.5   *Disposal of Waste Fluids*

Once the hydraulic fracturing is completed, fluid returns to the surface in a process known as "flowback". The US Environmental Protection Agency (EPA) estimates that fluids recovered range from 15–80% of the volume injected, depending on the site conditions. Therefore, each well will generate 1.5–16 million litres of flowback fluids which contain water, sand, methane, fracturing chemicals and contaminants released from the rock being fractured. These contaminants could include heavy metals, organic compounds and naturally occurring radioactive materials. Approximately 60% of the flowback fluids occur within the first four days, with the remaining 40% occurring in the next ten days.

The fluids that are not recovered remain underground and concerns have been expressed that these could become a source of contamination to underground aquifers in the future. Concerns have also been expressed as to the environmental risk as a result of either waste fluids disposal or a leak from the waste fluids storage facilities. The waste fluid from hydraulic fracturing can be managed in a variety of ways, including re-use for further hydraulic fracturing, but this is more practical in multi-stage hydraulic fracturing. However, the re-use may concentrate the contaminants in the fluid, making it harder to dispose of or to remove in water treatment plants. Waste fluid can be disposed of through injection into deep underground wells, it can be treated at local water treatment facilities to make it acceptable for returning to the environment provided it meets the water treatment standards, or it can be stored in tanks or deep lined pits. The size of these pits could be substantial: to accommodate up to 160 million litres of fluid from a multi-well pad would need a pit of 160 000 cubic metres per pad.

Underground injection is the primary disposal method for most shale gas projects worldwide, but whether this will be acceptable to the United Kingdom's relevant Environmental Agencies has yet to be determined. Where injection is not an option, new wastewater treatment facilities are being built in some parts of the world. The funding for building these plants varies, with some built using local taxpayers' money, with the gas companies paying for the volume of wastewater treated; in other cases the gas companies are paying for the construction of the plants while some are built under joint ventures with both the taxpayers and the gas companies funding the projects. The percentage of wastewater that is being recycled is increasing as companies become more adept at handling this waste and on-site treatment technologies become more readily available.

### 2.2.2.6   *Greenhouse Gas Emissions*

When evaluating the overall merit of any energy source many people look at the greenhouse gas emissions relative to the energy produced, *e.g.* the amount of greenhouse gases produced by burning coal to produce electricity. In the case of shale gas extraction, concerns have been expressed that there are significant fugitive emissions of methane associated with the process of shale gas extraction itself.

A study by Howarth *et al.* published in 2010 stated "Compared to coal, the (greenhouse gas) footprint of shale gas is at least 20% greater".[23] Fugitive emissions are losses of methane that occur between the well caps and the end user. These can be as a result of leakage around the site, leaks in distribution pipelines between the well and the end user, leakage from venting, *etc.* Most of these sources would be the same, whether it is a conventional gas well or a hydraulically fractured gas well. However, the fugitive emissions from the hydraulic fracturing process are of greater concern, as they have not been fully examined in detail and could potentially be very high.

After hydraulic fracturing but before the well is capped and the gas piped away for use, the well will be cleaned by flushing fluid or gas into the bore to remove debris and also flushing away produced methane that is not profitable to store or transport. In a study based on US Environmental Protection Agency data, academics from Cornell University have calculated that the venting of such emissions could mean that shale gas may be actually worse for global warming than coal (see Figure 5).[22]

Although coal releases more carbon dioxide per unit energy than methane, methane is a far more potent greenhouse gas – 72 times more powerful per unit mass than carbon dioxide over a 25-year period, falling to 25 times over a 100-year period. In the United Kingdom the Environment Agency believes that good practice can mitigate such fugitive emissions and is considering options to monitor air near the site to keep track of any leaks.

Venting in itself is a dangerous procedure, even if it were permitted in the UK by the Health and Safety Executive. Cuadrilla, which is one of the leading exploration companies operating at present in the UK, plans to flare the methane that is produced during testing prior to the production phase.[24]

In the US there are also concerns about methane emissions and the US EPA runs a voluntary program, EPA Natural Gas STAR, for companies adopting strategies to reduce methane emissions. These procedures are known by a variety of terms, including "the green flowback process" and "green completions."[25,26] To reduce the emissions, the gases and liquids brought to the surface during the completion processes are collected, filtered and then placed into the production pipelines and tanks instead of being dumped, vented or flared. The gas clean-up during a "green"

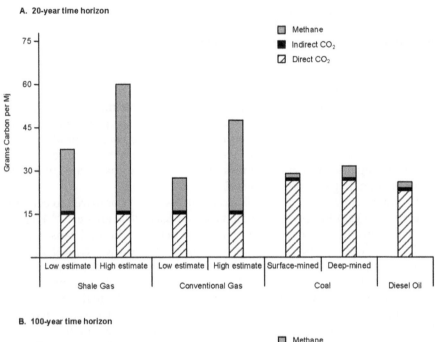

A. 20-year time horizon

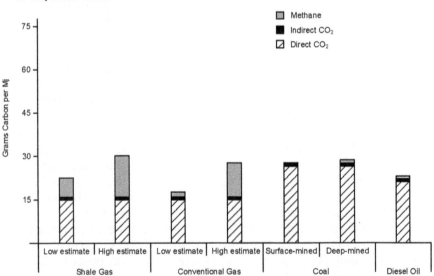

B. 100-year time horizon

**Figure 5**   Low and high estimates of lifecycle carbon dioxide equivalent from various fuel sources.[50]

completion is done with special temporary equipment at the well site and after a period of time (days) the gas and liquids being produced at the well are directed to permanent separators and tanks. The gas is then transported through piping and meters that are installed at the well site. Green completions methods do not involve complex technology and can be very cost effective. If this process for minimising fugitive emissions can be carried out cost-effectively in the US (where the payback period if these "Reduced Emissions Completions" are adopted is just a few months)[25] then the process would make even more financial sense in the UK where gas prices are considerably higher.

Another factor in favour of capturing methane instead of flaring is that flaring produces carbon dioxide (a greenhouse gas) as well as carbon monoxide, aromatic hydrocarbons and particulate matter emissions.

The establishment of fugitive and vented emissions of methane from hydraulic fracturing, flowback and its impact on greenhouse gases is still being debated. In addition to the study by Howarth *et al.*,[22] work has also been carried out by Jiang *et al.*[27] and Skone,[28] with each including an estimate of the methane emissions in their work.

In the case of Howarth *et al.*, they used five industry presentations, empirical data and lifetime emissions per well and provided figures of between 140 and 6800 thousand cubic metres of methane per well completion. They used a statistical uncertainty analysis to investigate different ratios of vented and flared gas.

Jiang used an uncertainty model rather than empirical data and estimated figures per flowback event of between 30 and 1470 thousand cubic metres of methane.

Skone again used figures per flowback event, with re-fracturing being assumed to be equivalent to a completion; their data source was not clearly identified but they cited fugitive emissions between 132 and 330 thousand cubic metres of methane.

The US EPA released new estimates of fugitive emissions and revised methodologies in 2011. They derived emission factors from four studies presented at Natural Gas STAR technology-transfer workshops.[29] Each study had a range of underlying individual measurements from three to over a thousand. The EPA background technical documents combine these studies to identify a figure of 260 thousand cubic metres of fugitive emissions per well completion.

The greenhouse gas impact associated with the fugitive methane gas emissions from the wells can be assumed to be similar to conventional vertical wells during the initial drilling stage, as are the levels of carbon dioxide associated with the machinery and equipment used in the drilling operation. The emissions associated with the horizontal drilling are, without more specific data being available, assumed to be the same as those emitted during vertical drilling. The California Environmental Protection Agency Air Resources Board assumes diesel fuel consumption in vertical drilling of 18.7 litres per metre drilled.[30] This figure would equate to an emission factor of $49 \text{ kgCO}_2 \text{ m}^{-1}$ of well drilling. The additional fuel required for the horizontal drilling is site-specific. However, if we assume it is similar to that used for the vertical drilling, assuming an additional horizontal drilling of a 1000 m bore could lead to an additional 49 tonnes of carbon dioxide compared to a conventional well with no horizontal drilling.

**Table 3**  Truck visits over lifetime of six well pads.[15]

| | Per well | | Per pad | |
|---|---|---|---|---|
| Purpose | Low | High | Low | High |
| Drill pad and road construction equipment | | | 10 | 45 |
| Drilling rig | | | 30 | 30 |
| Drilling fluid and materials | 25 | 50 | 150 | 300 |
| Drilling equipment (casing, drill pipe, *etc.*) | 25 | 50 | 150 | 300 |
| Completion rig | | | 15 | 15 |
| Completion fluid and materials | 10 | 20 | 60 | 120 |
| Completion equipment (pipe, wellhead) | 5 | 5 | 30 | 30 |
| Hydraulic fracture equipment (pump trucks, tanks) | | | 150 | 200 |
| Hydraulic fracture water | 400 | 600 | 2400 | 3600 |
| Hydraulic fracture sand | 20 | 25 | 120 | 150 |
| Flow back water removal | 200 | 300 | 1200 | 1800 |
| **Total** | | | **4315** | **6590** |
| *... of which, associated with fracturing process:* | | | *3870* | *5750* |
| | | | *90%* | *87%* |

The emissions from the hydraulic fracturing process, which uses more fuel than a conventional well, relate to emissions from the fuel used by the high-pressure pumps required. New York State reports that emissions from these pumps, based on an average fuel use, for hydraulic fracturing of eight horizontally drilled wells in the Marcellus Shale used a total of 110 000 litres of diesel fuel, producing 295 tonnes of carbon dioxide per well.[31]

As with all industrial sites, during the construction and hydraulic fracturing phases there will be heavy road traffic moving in and out of the site. New York State (2009)[15] provides estimates of truck visits to a typical site. These are summarised in Table 3, giving trips per well and per well pad (based on six wells per pad). This suggests a total number of truck visits of between 4300 and 6600 of which about 90% are associated with the hydraulic fracturing operation.

These truck visits will also have an adverse impact on greenhouse gas emissions.

During the production phase the amount of traffic will be minimal, but once the site is completed there will again be considerable heavy traffic movements as equipment is removed and the site returned to its original use.

The fossil fuel emissions of shale gas *versus* conventional gas per well are given in Table 4 which, in addition to the emissions associated with the horizontal drilling and the hydraulic fracturing, also takes into account the transport of water to and from the site and the wastewater treatment. Water UK estimated the emissions to the atmosphere at the wastewater treatment plant to be 0.406 tonnes of carbon dioxide per thousand cubic metres of water treated.[32]

### 2.2.2.7  Road Traffic
In the UK many of the shale gas sites will be in rural locations served by minor roads which are often narrow and of relatively light road construction

**Table 4**  Key additional fossil fuel combustion emissions associated with shale gas extraction.

| Process | Emissions (tCO$_2$) | Assumptions |
|---|---|---|
| Horizontal drilling[a] | 15–75 | Horizontal drilling of 300–1500 m; 18.6 litres diesel used per metre drilled. |
| Hydraulic fracturing[b] | 295 | Based on average fuel usage for hydraulic fracturing on eight horizontally drilled wells in the Marcellus Shale. The total fuel use given is 109 777 litres of diesel fuel. |
| Hydraulic fracturing chemical production[c] | – | Unknown. |
| Transportation of water[d] | 26.2–40.8 | Based on HGV emission factor of 983.11 gCO$_2$ km$^{-1}$ and 60 km round trip. |
| Wastewater transportation[d] | 11.8–17.9 | Based on HGV emission factor of 983.11 gCO$_2$ km$^{-1}$ and 60 km round trip. |
| Wastewater treatment[e] | 0.33–9.4 | Based on 15–80% recovery of 9–29 million litres of water that is required per fracturing process and emission factor 0.406 tCO$_2$ ML$^{-1}$ treated. |
| **Total per well** | **348–438** | **Based on single fracturing process.** |

[a]Fuel consumption from: *ALL Consulting* (2008). Emission factor from *DUKES* (2010).[36]
[b]Cited from ALL Consulting, Horizontally Drilled/High-volume Hydraulically Fractured Wells Air Emissions Data, August 2009, Table 11, p. 10 by New York State (2009).[30] Emission factor from *DUKES* (2010).[36]
[c]A further potential source of additional emissions may be from the production of chemical used in the fracturing process. However, the level of these emissions is difficult to ascertain as: conventional wells may also include various chemicals in drilling mud, so claiming shale creates additional emissions *via* this route is problematic; and life cycle analysis data for these chemicals is highly specialised and is not typically publically available.
[d]Emission factor from NAEI (2010).[46] Truck numbers from Table 3.
[e]Emission factor from Water UK – Towards sustainability (2006).[31]

to reflect the low numbers of traffic movements and minimal use by heavy vehicles. The roads are likely to pass through small villages and towns which are not accustomed to industrial developments. The high number of large lorries visiting the shale gas sites during the construction phase will subsequently result in deterioration of the road structure as well as generating dust and noise associated with the traffic movements. This is likely to be an area of serious concern to the local residents along the route the traffic will take and will result in local objections at the planning approval stage. Once the sites obtain the necessary approval, the operators of the sites could encounter traffic movement issues as local residents and the farming community compete for road space with the construction traffic which could lead to conflict between the parties. This is an issue that the shale gas companies need to address at the early planning stage to minimise disruption and to work with the local community to prevent adverse public relation issues which could impact on future projects.

## 2.3   Production

The completion of drilling and hydraulic fracturing marks the start of the production phase of the wells. A production well head is put in place to collect and transfer the gas for subsequent processing, either for utilising on site for electricity production; for liquefaction and transport off site; or for piping into the gas distribution network. Production from one well can commence while other wells are being drilled and fractured.

In terms of production volumes, indicative figures for long-term production for a single Marcellus well in New York State are as follows:[15]

Year 1: Initial rate of 79 000 m$^3$ per day, declining to 25 500 m$^3$ per day
Years 2 to 4: 25 500 m$^3$ per day, declining to 15 600 m$^3$ per day
Years 5 to 10: 15 600 m$^3$ per day, declining to 6400 m$^3$ per day
Years 11 onwards: 6400 m$^3$ per day, declining at a rate of 3% per annum

As can be seen, after five years the volume of gas produced drops dramatically and it is at that point that the operator may decide to re-fracture the well to extend its life. Re-fracturing can take place more than once. It is anticipated that wells in the UK will follow a similar pattern of production drop-off over time.

*2.3.1   Decommissioning and Plugging Off of Wells.*   When the productive life of a well is over, or if wells prove uneconomic to exploit, the correct procedures need to be put in place to ensure that the wells are correctly plugged and abandoned. Proper plugging is critical to protect the groundwater aquifers, surface water and soil. Well plugging involves removal of the well head and the removal of the downhole equipment. Uncemented casings in critical areas must either be pulled up or perforated. Cement must then be placed in the wellbore to seal the bore or squeezed through the perforations of the casings if they remain in place to seal between the casings, the rock formations and to fill the bore. This procedure occurs at intervals dictated by the relevant regulatory authority to ensure a seal between hydrocarbon- and water-bearing zones. As an example of how an individual American state's regulations have evolved in specific detail, California's plugging regulations require cement plugs to be placed in the following locations: a 200-foot plug straddling the surface casing shoe; a plug across oil- and gas-bearing strata that extends 100 feet above the strata; a plug extending from 50 feet below to 50 feet above the base of the water-bearing strata; and a 50-foot plug at the surface of the wellbore.[33] In the UK, the Environmental Agency in England, the Scottish Environmental Protection Agency or the Environmental Agencies of Wales and Northern Ireland will require the operators to put in place procedures to ensure that there is no future release or escape of shale gas or waste water either into the environment or into groundwater-bearing strata.

## 3 UK Estimates of Shale Gas

In the UK, the government Department of Climate Change (DECC) defines the estimates of gas in the following terms:

*Proven:* reserves which, on the evidence available, are virtually certain to be technically and commercially producible, *i.e.* have a better than 90% chance of being produced.

*Probable:* reserves which are not yet proven, but which are estimated to have a better than 50% chance of being technically and commercially producible.

*Possible:* reserves which at present cannot be regarded as probable, but which are estimated to have a significant but less than 50% chance of being technically and commercially producible.

Terminology referring to smaller physical scales is also used. Gas Initially In Place (GIIP), or Gas In Place (GIP) for the remainder if production has commenced, refers to the total gas resource that is present in a reservoir or gas field and is a resource rather than a reserve measure. Estimated Ultimately Recoverable (EUR) refers to a given well or field over its lifetime and accounts for its production to date and anticipated recovery. This measure is closer in sense to a reserve.

The British Geographical Survey (BGS) in association with the Department of Energy and Climate Change (DECC) completed an estimate for the resources (gas in place) of shale gas in part of central Britain in an area between Wrexham and Blackpool in the west, and Nottingham and Scarborough in the east.[34] The estimate is in the form of a range in order to reflect geological uncertainty. The lower limit of the range is 822 trillion cubic feet (tcf) and an upper limit of 2281 tcf, but the central estimate for the resource is 1329 tcf (see Figure 6).

This shale gas estimate is a resource figure and so represents the gas that is thought to be present, but not the gas that it might be possible to extract.

Prediction of reserves then needs to be considered against what can be economically recovered, which of course depends on the price of gas within the market it is operating in. The volumes of gas that can be actually recovered against the predicted reserves are very difficult to determine until exploratory wells have been drilled. While figures of between 10 and 20% of predicted reserves could be recoverable, in some instances the gas actually recovered has been very much lower.

In terms of estimates from individual licence areas that have already been allocated in the UK, four companies have provided estimates of reserves but to date only one company, Cuadrilla Resources, has carried out exploratory drilling. This exploration of the commercial shale gas extraction has taken place in the Bowland Shales in Lancashire. The company's UK Petroleum

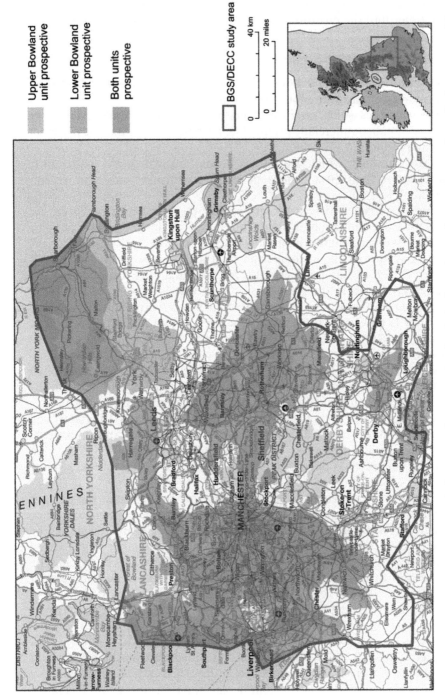

**Figure 6**  Location of the BGS/DECC shale gas study area, central Britain (contains Ordnance Survey data, © Crown copyright and database, 2013).[51]

Exploration and Production Licence (PEDL) was granted in September 2008. (A PEDL is awarded for six years initially on the basis of the applicant demonstrating technical and financial competence and an awareness of the environmental issues. The licensee is also required to demonstrate that they have obtained access rights from relevant landowners and complied with other statutory planning laws).

Work commenced on site, drilling the first test well in August 2010 at Preese Hall Farm followed by a second at Grange Hill Farm later that same year. Work on the third, near the village of Banks, commenced in August 2011. Based on the initial exploration, Cuadrilla Resources announced its first estimate of the volume of shale gas within its licence area on 21st September 2011. It estimates the total gas initially in place to be 5660 billion cubic metres (bcm) of gas; if it were able to extract 20% of the gas this would equate to 1132 bcm of recoverable gas.

Island Gas Limited (IGL) and its subsidiary company IGas Resources PLC operate in the North of England and North Wales and, using borehole logs, they have identified shale deposits extending over 1195 km$^2$ with an average depth of 250 m. In October 2010 they gave estimates of gas initially in place as between 2.5 bcm and 131 bcm, with a risk factor of 50% for their North Wales licence area. No estimates of recoverable gas have been given and they intend to conduct further exploratory work in the future to fully understand the potential of these deposits.[35]

Eden Energy Ltd commissioned an independent expert report from RPS Group plc, a multinational energy resources and environmental consultancy company, in respect of prospective gas reserves in the 806 km$^2$ of the seven PEDLs in South Wales. They have estimated that the proven reserves of gas initially in place are 968 bcm, with recoverable volume of 362 bcm.

Dart Energy, which took over Composite Energy in 2011, has indicated shale potential in addition to coal-bed methane extraction within its licence area of north west England and Wales with an estimate of total gas initially in place of 34 bcm.[36]

In April 2012, Dart Energy completed the acquisition of Greenpark Energy's 17 UK licences. Of these licences, PEDL159 is the furthest developed and is situated in Scotland. Eight wells were drilled in the area by Greenpark Energy, with three of these wells being pilot production wells, which demonstrated good production rates of natural gas from the coal seams. The remaining five wells were exploration wells; these delineated seam thickness, gas contents, gas saturations and permeability trends in the area. Since the acquisition, Dart Energy has been assessing the licence area from both a sub-surface and surface perspective.

In total, Dart Energy now have four licences in Scotland and a further 28 licences in rest of the UK.

Shale gas production in the UK is still some years off, with production unlikely to start before 2015–2016. Estimates of production are still very difficult to obtain, but Cuadrilla have estimated that commercial development of the resource would provide an annual average of 0.7 to 2.8 bcm of gas. With cumulative estimates of between 19.7 and 76.7 bcm total gas production by 2040 based on the well construction figures for the Bowland shale shown in Table 5. This represents between 1.7 and 6.8% of the estimated 1132 bcm recoverable resource.

The figures for possible maximum annual production of shale gas from projections relating to Cuadrilla and the Bowland shales of 4.90 bcm shown in Table 6 need to be set against the annual UK gas consumption in 2010 of 91 bcm.[37]

**Table 5**   Commercialisation scenarios for Cuadrilla Resources, Bowland Shale.[47]

| Year | Well construction low | Well construction central | Well construction high |
|---|---|---|---|
| 2013 | 20 | 20 | 20 |
| 2014 | 30 | 30 | 30 |
| 2015 | 40 | 40 | 40 |
| 2016 | 40 | 60 | 60 |
| 2017 | 40 | 60 | 60 |
| 2018 | 20 | 60 | 60 |
| 2019 | | 60 | 60 |
| 2020 | | 40 | 60 |
| 2021 | | 30 | 60 |
| 2022 | | | 60 |
| 2023 | | | 60 |
| 2024 | | | 60 |
| 2025 | | | 60 |
| 2026 | | | 60 |
| 2027 | | | 40 |
| 2028 | | | 20 |
| Total wells | 190 | 400 | 810 |
| Wells per pad | 10 | 10 | 10 |
| Total pads | 20 | 40 | 80 |
| Duration of activity (years) | 6 | 9 | 16 |
| Peak activity (wells drilled per year) | 40 | 60 | 60 |

**Table 6**   Estimated gas production statistics under Cuadrilla commercialisation scenarios (2014–2040).[46]

| | Low | Medium | High |
|---|---|---|---|
| Cumulative Production (bcm) | 19.7 | 40.3 | 76.7 |
| Cumulative as a percentage of estimated recoverable resource (1132 bcm) | 1.7% | 3.6% | 6.8% |
| Average annual production (bcm) | 0.73 | 1.49 | 2.84 |
| Average annual production as a percentage of UK consumption in 2010 (91 bcm) | 0.8% | 1.7% | 3.2% |
| Minimum production in a single year (bcm) | 0.29 | 0.58 | 0.58 |
| Maximum production in a single year (bcm) | 2.12 | 3.57 | 4.90 |

## 3.1    Resource Requirements

In terms of resource requirements, Table 7 provides information on the activities and resources required for development of shale gas pads in the US, where the resources for the vertical bore are separated from the resources associated with the horizontal bore and the hydraulic fracturing.[15]

Based on the information from similar operations in the US shown in Table 7 and using the commercialisation scenarios from Table 5, it is possible to estimate the resource requirement under Cuadrilla Resources' various scenarios (see Table 8).

As can be seen from Table 6, the high scenario involving the development of 810 wells will provide a cumulative volume of shale gas of 76.6 bcm over a 25-year timescale, with an annual production of 2.84 bcm. Therefore, to have a significant impact on the annual gas consumption in the UK, significantly more wells need to be developed in the UK as a whole.

To achieve a 10% production of UK annual gas consumption of 91 bcm would require of the order of 250 wells to be in production at any one time and, bearing in mind the productivity of individual wells over time decreases

**Table 7**   Summary of resources, pre-production.[46]

|  | Activity | Six well pads drilled vertically to 2000 m and laterally to 1200 m | |
|---|---|---|---|
|  |  | Low | High |
| **Construction** | Well pad area (ha) | 1.5 | 2 |
| **Drilling** | Wells | 6 | |
|  | Cuttings volume (m³) | 827 | |
| **Hydraulic Fracturing** | Water volume (m³) | 54 000 | 174 000 |
|  | Flowback fluid volume (m³) | 7920 | 137 280 |
| **Surface Activity** | Total duration of surface activities pre-production (days) | 500 | 1500 |
|  | Total truck visits | 4315 | 6590 |
| **Re-fracturing Process** | Water volume (m³) | 27 000 | 87 000 |
| Assuming an average of 50% of wells re-fractured only once | Fracturing chemicals volume, @ 2% (m³) | 540 | 1740 |
|  | Flowback fluid volume (m³) | 3960 | 68 640 |
|  | Total duration of surface activities for re-fracturing (days) | 200 | 490 |
|  | Total truck visits for re-fracturing | 2010 | 2975 |
| **Total for 50% re-fracturing** | Well pad area (ha) | 1.5 | 2 |
|  | Wells | 6 | |
|  | Cuttings volume (m³) | 827 | |
|  | Water volume (m³) | 81 000 | 261 000 |
|  | Fracturing chemicals volume, @ 2% (m³) | 1620 | 5220 |
|  | Flowback fluid volume (m³) | 11 880 | 205 920 |
|  | Total duration of surface activities pre production (days) | 700 | 1990 |
|  | Total truck visits | 6325 | 9565 |

**Table 8**  Resource requirements per well under Cuadrilla development scenarios.[48]

|                                        | Resources use per well        |
|----------------------------------------|-------------------------------|
| Wells                                  | 1                             |
| Well pads                              | 0.1 (*i.e.* 10 wells per pad) |
| Well pad area (ha)                     | 0.7                           |
| Water volume ($m^3$)                   | 8399                          |
| Fracturing chemicals volume ($m^3$)    | 3.7                           |
| Cuttings volume ($m^3$)                | 138                           |

**Incorporating data from Table 7**

|                                                            | Low Estimate | High Estimate |
|------------------------------------------------------------|--------------|---------------|
| Flowback fluid volume ($m^3$)                              | 1232         | 6627          |
| Total duration of surface activities pre production (days) | 83           | 250           |
| Total truck visits                                         | 719          | 1098          |

**Table 9**  Resource requirements to deliver 9 bcm per year for 20 years.

|                                      | Assuming no re-fracturing | | Assuming a single re-fracturing on 50% of wells (delivering an assumed 25% increase in productivity for those wells) | |
|--------------------------------------|------------|-----------|------------|-----------|
| Wells                                | 2970       |           | 2592       |           |
| Well pads                            | 297        |           | 259        |           |
| Cuttings volume ($m^3$)              | 409 365    |           | 357 264    |           |
| Water volume ($m^3$)                 | 24 945 030 |           | 32 655 312 |           |
| Fracturing chemicals volume ($m^3$)  | 10 989     |           | 14 386     |           |
|                                      | Low        | High      | Low        | High      |
| Flowback fluid volume ($m^3$)        | 3 658 604  | 19 680 768| 4 789 446  | 25 763 915|
| Total truck visits                   | 2 135 925  | 3 262 050 | 2 732 400  | 4 132 080 |

rapidly, the figure in reality would have to be considerably greater to take account of the decline in production, with new wells having to come online or existing wells re-fractured. Over a 20-year period, between 2600 and 3000 wells would need to be developed to deliver a sustained annual output of 9 bcm. This equates to between 260 and 300 well pads, assuming that 10 wells can be drilled and hydraulically fractured from each well pad.

To put this into context, the DECC in 2010 identified that only 2000 wells had been drilled in total in the UK, with about 25 onshore wells drilled per year in the UK in the last decade.

Based on the information from Cuadrilla and the US, Table 9 provides the total resources required in the UK for production of 9 bcm of shale gas, which represents 10% of the annual UK gas consumption in 2010.

## 4  Existing Natural Gas Supplies in the UK

The UK at present gets its gas from a variety of sources, including the UK Continental Shelf (UKCS) where production has been in decline since 2000, and in 2012 it was 38% (43 bcm) of the level produced in 2000 (114 bcm). Since 2000 the rate has declined on average by about 8% per annum, but the decline varies each year depending on operational issues and the price of gas on the world market, which determines the viability of marginal production. In 2012 production was 14% lower than in 2011. This was largely due to operational issues where a leak on the Elgin platform in the North Sea in March 2012 reduced production for the rest of 2012.

The UK imports natural gas by pipelines from Norway, Belgium and the Netherlands, and liquefied natural gas (LNG) *via* ships. The UK has been a net importer of gas since 2004, with net imports in 2012 accounting for 47% of supply. In 2012 the UK imported approximately 50 bcm. In 2012 total LNG imports to the UK *via* the four import terminals at Dragon (Milford Haven), Isle of Grain, South Hook (Milford Haven) and Teesside GasPort were approximately 13.5 bcm (see Figure 7).[36]

With the two interconnecting pipes to mainland Europe, the UK gas network is integrated with the wider continental gas network, with gas pipelines extending into Russia, the Middle East and North Africa. These pipelines enable gas to be moved around from various locations internationally and enable the gas producers to buy and sell gas on the international market without the need to liquefy and ship it. This reduces costs and therefore ensures that the UK can not only import gas *via* pipelines from a wide geographical area but also allows UK producers to export UKCS gas to Europe.

It is against the background of international trade, which contrasts with the situation in the US that any shale gas production in the United Kingdom needs to be considered.

## 5  World Energy Environment: Is Shale likely to be a Game Changer?

In North America, which has very limited export markets for the shale gas it produces, with pipelines extending into Mexico and no LNG export terminals, any shale gas produced is used predominantly within the US. This led to a dramatic reduction in the cost of gas within the US, with gas prices today (in 2013) at about a third of the peak energy price in 2008 (see Figure 8).

The US is now in the process of converting liquefied natural gas (LNG) import terminals into LNG export terminals. A recent deal between the BG Group and the US energy firm Cheniere Energy Inc. may allow 3.5 million tonnes per year of LNG to be exported from the US to Europe from 2015 onwards at prices lower than Asian or European gas.[38] It will be interesting to see whether the exporting of LNG from the US to the world market will impact on the price of shale gas within the US and, more importantly for the

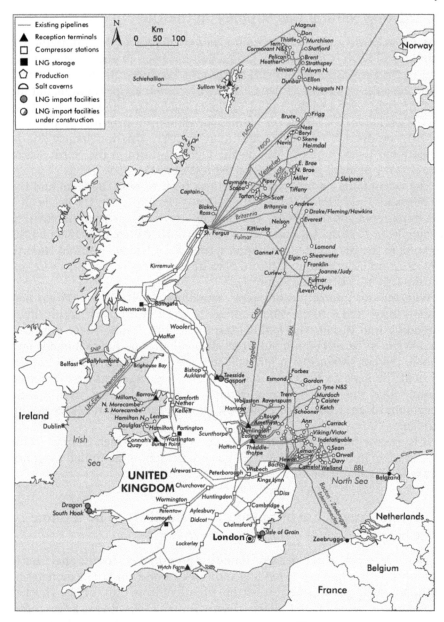

**Figure 7**   The National Gas Transmission System 2012.[36]

UK, whether the price of gas falls. This will provide, for the first time, direct competition between US-produced shale gas and locally produced natural gas in Europe.

As of 1st October 2013, the US Department of Energy (DOE) has approved only four applications for permits to export LNG to non-free trade agreement

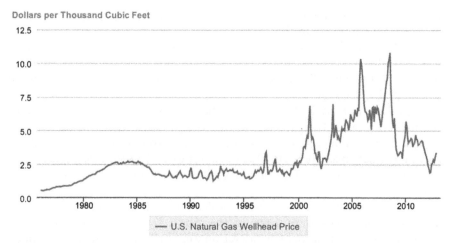

**Figure 8**   United States gas wellhead price.[52]

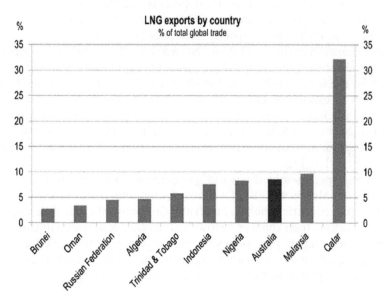

**Figure 9**   LNG Exports by country, 2012.[39]

nations. There are currently 21 pending applications, covering 19 facilities where US businesses are seeking to build and operate terminals to process LNG for export.[39] The most recent decision was made in September 2013, when Dominion Resources Inc. received approval for the Cove Point Terminal on the Maryland shore of Chesapeake Bay. To date, the DOE has authorised 6.37 bcf of LNG from the plant to be sold overseas. Delays in the approval process stem from the debate within the US administration on how much natural gas should be exported without raising the natural gas prices and reducing supplies in the US.

As well as supplies of shale gas from America using LNG export terminals, Australia also plans to expand in the export market for LNG. It now accounts for 9% of the globally traded LNG and production is set to ramp up substantially (see Figure 9).

Investment in capacity to export LNG has been the key driver of Australia's recent boom, with 7 of the world's 12 largest LNG plants being built in Australia. Investment in Australia accounts for around two thirds of all the current global investment in the LNG industry.[40] Exports are set to rise sharply and government forecasts suggest a rise of over 300% between 2015 and 2020.

Based on Australian government projections, Australia will become the second-largest exporter of LNG in the world by 2016. If all the major investments projects under consideration were to go ahead, Australia stands to become the world's largest exporter of LNG by 2020, overtaking Qatar. This is without the potential shale gas and other unconventional gas sources being heavily exploited. These developments are in their early stages but could become a significant source of the country's gas production, especially given that the LNG export capacity is already being built.

This expansion in the Australian LNG market will be in direct competition with the expansion of the US shale gas export proposals and will be competing for the same markets worldwide. It will be interesting to see how the expansion in exports of LNG affects the development of shale gas in other areas and the impact that this has on world gas prices. Should world prices reflect prices in the US, then it is likely that that there will be a slowing down of future exploration in countries where regulation and operating conditions are more challenging.

If, due to increases in the world gas demand (not only from developing countries but also as a result of a change in national policy relating to power generation in countries such as Japan and Germany), there is little impact on world gas prices, then shale gas extraction in other countries will continue to be developed. Japan and Germany are expanding their gas-fired power station programmes as a result of trying to reduce their nuclear generation capacity in response to the Fukushima nuclear accident that occurred in 2011.

## 6  Regulation

Regulation in the UK is controlled by a variety of national and local government departments with a variety of different responsibilities.

In the UK, the environmental considerations are controlled by the following organisations: the Environmental Agency (EA) in England, the Scottish Environmental Protection Agency (SEPA) and the Environmental Agencies of Wales and Northern Ireland. These agencies are responsible for the protection of the environment and for the exploration and extraction of unconventional gases for which they perform various roles. The agencies provide advice to central government and regulate operators to ensure that the environment is protected. They also advise government on the sustainability of these sources of natural gas and their associated extraction technologies. The agencies ensure that the exploration and development of unconventional gases is regulated effectively to manage risks to surface water and groundwater resources. In addition, they are responsible for granting any necessary environmental permits and have powers to serve notices, where required, to protect the local environment.[41] This is done by applying a proportionate and risk-based approach to preventing pollution and protecting the environment. The EA is responsible for regulating water abstraction and regulating any discharge associated with the extraction process and is also a statutory consultee in the planning process and will provide advice to local authorities on individual gas extraction sites.

The Department of Energy and Climate Change (DECC) administers the licensing system under the *Petroleum Act 1998*, which authorises each particular drilling as a development activity. DECC provides up-to-date information on its website relating to the various *Approved Codes of Practice*, official guidance and referenced codes, standards, *etc.,* which need to be taken into account when proposing to develop a shale gas deposit.[42]

The Planning Authority (generally the local authority) deals with the planning application that is required for each site. The councillors who are responsible for approving these planning applications are locally elected representatives and as such are keenly aware of the local issues and the impact that these decisions will have on the local electorate.

The Health and Safety Executive (HSE) regulates the safety aspects of the work, which contributes to the mitigation of the environmental risk. In particular, they are responsible for regulating the appropriate design, construction and continued integrity of any gas well. In addition, the rest of the requirements under the *Health and Safety at Work Act 1974* will also need to be considered when proposing and undertaking any works.

The Institution of Gas Engineers and Managers (IGEM) has published a guidance document, *IGEM/G/101* on *Onshore Natural Gas Extraction and Route to Use,* which provides an overview of the requirements in the UK.[43]

## 7 Predictions are Hard, Especially about the Future

Shale gas has not yet been developed on a commercial basis in the UK, which inevitably means that one has to make a series of assumptions about how successful the shale gas industry will become and the effect of shale gas on energy prices within the UK economic environment.

As many experts have identified, there are a number of barriers and issues about UK shale gas that are different from the situation applicable in the USA. These include different geology, different regulation and a much greater density of population. The experience of other European countries also highlights that, even within Europe, opinions relating to shale gas are divided. France, for example, currently has a moratorium on shale gas, whilst in Poland the government is actively supporting shale gas development although the industry faces various problems here that are affecting its development.

The US Energy Information Administration estimated in 2011 that Poland had possible reserves of 5.3 trillion cubic metres of shale gas, the biggest in Europe. More recently, the Polish Geographical Institute has been more conservative, with estimates of between 346 and 768 billion cubic metres of shale gas.[44] However, there is a fear that Poland is moving to tax and regulate the shale gas industry more and this uncertainty, coupled with the complicated geology, which has produced some disappointing early results from test wells, has lowered the enthusiasm of international exploration companies.

So the development of shale gas across the UK and the rest of Europe is a mixed and complicated prospect that is difficult to predict, both in terms of the volumes likely to be extracted and the impact this will have on overall energy prices.

Worldwide predictions get even more difficult. The US Energy Information Administration published figures in 2013 of the technically recoverable shale gas resources outside of the United States which identified China as having reserves of 31.55 trillion cubic metres and the top ten countries, which include the US, as having reserves of 206 trillion cubic metres (see Table 10).[48] However, how much of these reserves are economically recoverable is another matter. Key positive above-the-ground advantages in the US and Canada that may not apply in other locations include private ownership of sub-surface rights that provide a strong incentive for development. These include the availability of many independent operators and supporting contractors with critical expertise and suitable drilling rigs and the pre-existing gathering pipeline infrastructure, together with the availability of water resources for use in the hydraulic fracturing.

**Table 10**  Top ten countries with technically recoverable shale gas resources.[49]

| Rank | Country | Shale gas reserves (trillion cubic metres) |
|------|---------|--------------------------------------------|
| 1 | China | 31.55 |
| 2 | Argentina | 22.7 |
| 3 | Algeria | 20 |
| 4 | USA[a] | 18.82  (32.85) |
| 5 | Canada | 16.22 |
| 6 | Mexico | 15.42 |
| 7 | Australia | 12.37 |
| 8 | South Africa | 11.04 |
| 9 | Russia | 8.06 |
| 10 | Brazil | 6.93 |
| | **World Total** | **206.56**  (220.6) |

[a]Energy Information Agency estimates used for ranking order. Advanced Resources International estimates in parentheses.

Given the variation across the world's shale formations in both geology and above-ground conditions, the extent to which global technically recoverable shale resources will prove to be economically recoverable is not yet clear.

The impact that shale gas will have on the UK gas supply situation and the price to gas customers is impossible to determine due to the large amount of uncertainty.

Lord Browne, one of the most powerful energy figures in the UK and Chairman of Cuadrilla, the leading shale gas company in the UK, told an audience at the Grantham Research Institute at the London School of Economics (LSE) on the 27th November 2013,

"I don't know what the contribution of shale gas will be to the energy mix of the UK. We need to drill probably 10 to 12 wells and test them and it needs to be done as quickly as possible.

"We are part of a well-connected European market and unless it is a gigantic amount of gas, it is not going to have a material impact on price."[45]

These two sentences effectively sum up the situation within the UK at this time (December 2013).

## Disclaimer

The views expressed in this chapter are those of the author and do not necessarily reflect the position or policy of the Institution of Gas Engineers and Managers.

## References

1. R. C. Selley, UK shale gas: The story so far, *Mar. Pet. Geol.*, 2012, **31**(1), 100–109; http://www1.gly.bris.ac.uk/bumps/secure/POTW/Selley_2012.pdf (last accessed 16/01/2014).
2. D. Donohue, N. Anstey and N. Morrill, Shale Gas in the Southern Central Area of New York State, US Dept, *Energy, Morgantown*, 1981, vol. 4, 578.
3. R. E. Zielinski and R. D. McIver, Resources and Exploration Assessment of the Oil and Gas Potential in the Devonian Shale Gas of the Appalachian Basin. US Dept. of Energy, Morgantown, 1998.
4. R. C. Selley, British shale gas potential scrutinized, *Oil Gas J.*, 1987, **15**, 62–64.
5. J. B. Curtis, Fractured shale-gas systems, *Bull. Am. Assoc. Pet. Geol.*, 2002, **86**, 1921–1938.
6. A. M. Martini, L. M. Walter, J. M. Budai, T. C. W. Ku, C. J. Kaiser and M. Schoell, Genetic and temporal relations between formation waters and biogenic methane. Upper Devonian Antrim Shale, Michigan Basin, USA, *Geochim. Cosmochem. Acta.*, 1998, **62**(10), 1699–1720.
7. West Virginia Surface Owners' Rights Organisation, Why multiple horizontal wells from centralized well pads should be used for the

    Marcellus Shale, 2012; http://www.wvsoro.org/resources/marcellus/
    horiz_drilling.html (last accessed 16/01/2014).
 8. Baker Hughes. *Rig Count Data*, 2013; http://phx.corporate-ir.net/phoenix.
    zhtml?c = 79687&p = irol-rigcountsoverview (last accessed 16/01/2014).
 9. U.S. Energy Information Administration. *Independent Statistics and An-
    alysis 2013: Official Shale Gas production data from Form EIA-895*, 2013;
    http://www.eia.gov/ (last accessed 16/01/2014).
10. The Petroleum Economist Limited in association with BG Group. *World
    LNG Factbook. 2012 Edition*. The Petroleum Economist Limited, London,
    2012.
11. N. J. P. Smith, Unconventional hydrocarbons: changing exploration
    strategies, *Earthwise*, 1995, 7, 14–15.
12. G. Swann and J. Munns, The Hydrocarbon Prospectivity of Britain's
    Onshore Basins, DTI, 2003, 17.
13. R. C. Selley, *UK shale-gas resources, Pet. Geol.: North-West Europe Global
    Perspect., Proc. 6th Petroleum Geology Conference*, 2005, Geological Soci-
    ety, London, 2005, 707–714.
14. Department for Energy and Climate Change, *Oil and Gas: Onshore Ex-
    ploration and Production*, 2013; https://www.gov.uk/oil-and-gas-onshore-
    exploration-and-production#resumption-of-shale-gas-exploration     (last
    accessed 16/01/2014).
15. Department of Environmental Conservation, *Well Permit Issuance for
    Horizontal Drilling and High-Volume Hydraulic Fracturing in the Marcellus
    Shale and Other Low-Permeability Gas Reservoirs*, 2011; http://www.dec.
    ny.gov/energy/75370.html (last accessed 16/01/2014).
16. J. Broderick, K. Anderson, R. Wood, P. Gilbert and M. Sharmina, *Shale
    Gas: An Updated Assessment of Environmental and Climate Change Impacts*,
    A report commissioned by The Co-operative and undertaken by re-
    searchers at the Tyndall Centre, University of Manchester, 2011; http://
    www.co-operative.coop/Corporate/Fracking/Shale%20gas%20update
    %20-%20full%20report.pdf (last accessed 16/01/2014).
17. UK Parliament, *Shale Gas – Energy and Climate Change*, 2011; http://www.
    publications.parliament.uk/pa/cm201012/cmselect/cmenergy/795/
    795we09.htm (last accessed 3rd February 2014).
18. House of Commons Committee of Public Accounts, *Ofwat: Meeting the
    Demands for Water*, 2007; [http://www.publications.parliament.uk/pa/
    cm200607/cmselect/cmpubacc/286/286.pdf (last accessed 16/01/2014).
19. FracFocus. *What Chemicals are Used*, 2014; http://fracfocus.org/chemical-
    use/what-chemicals-are-used (last accessed 3rd February 2014).
20. FracFocus. *What Chemicals are Used*, 2014; http://fracfocus.org/chemical-
    use/what-chemicals-are-used (last accessed 3rd February 2014).
21. C. J. De Pater and S. Baisch, *Geomechanical Study of Bowland Shale
    Seismicity. Synthesis Report*, 2011; http://www.cuadrillaresources.com/
    wp-content/uploads/2012/02/Geomechanical-Study-of-Bowland-Shale-
    Seismicity_02-11-11.pdf (last accessed 16/01/2014).

22. C. Frohlich, *Study Finds Correlation Between Injection Wells and Small Earthquakes*, 2012; http://www.utexas.edu/news/2012/08/06/correlation-injection-wells-small-earthquakes/ (last accessed 3rd February 2014).

23. R. Howarth, R. Santoro, A. Ingraffee, A commentary on "The greenhouse-gas footprint of natural gas in shale formations", *Climate Change*, 2012, **113**(2), 525–535; http://download.springer.com/static/pdf/112/art%253A10.1007%252Fs10584-011-0333-0.pdf?auth66 = 1390571575_f9a31aef677463804c72b217e82e970d&ext = .pdf (last accessed 16/01/2014).

24. Institution of Gas Engineers and Managers, *Shale Gas A UK Energy Miracle? An IGEM Report*, 2011; http://www.igem.org.uk/media/107958/igem-shale_gas-a_uk_energy_miracle-september_2011.pdf (last accessed 16/01/2014).

25. R. Guerra, *Green Flowback Process, Integrated Production Services*, 2005; http://www.epa.gov/gasstar/documents/workshops/houston-2005/green_flowback.pdf (last accessed 16/01/2014).

26. U.S. Environmental Protection Agency, *Lessons Learned from Natural Gas STAR Partners. Reduced Emissions Completions for Hydraulically Fractured Natural Gas Well*, 2011; www.epa.gov/gasstar/documents/reduced_emissions_completions.pdf (last accessed 16/01/2014).

27. J. M. Jiang, W. M. Griffin, C. Hendrickson, P. Jaramillo, J. VanBriesen and A. Venkatesh, Life cycle greenhouse gas emissions of Marcellus shale gas, *Environ. Res. Lett.*. 2011, **6**(3), 034014; http://iopscience.iop.org/1748-9326/6/3/034014/pdf/1748-9326_6_3_034014.pdf (last accessed 16/01/2014).

28. T. J. Skone, *Life Cycle Greenhouse Gas Inventory of Natural Gas Extraction, Delivery and Electricity Production*, U.S. Department of Energy, 2011; http://www.netl.doe.gov/energy-analyses/pubs/NG-GHG-LCI.pdf (last accessed 16/01/2014).

29. U.S. Environmental Protection Agency. *U.S. Greenhouse Gas Emissions and Sinks: 1990-2011*, 2011; www.epa.gov/climatechange/ghgemissions/usinventoryreport.html (last accessed 16/01/2014).

30. California Department of Conservation Oil and Gas Division. *Monthly Consumption Estimates of Natural Gas for Lease Fuel according to each Operator and Gas Field in California*, 2001; www.arb.ca.gov/ei/areasrc/ccosmeth/att_l_fuel_combustion_for_petroleum_production.doc (last accessed 16/01/2014).

31. ALL Consulting. *Evaluating the Environmental Implications of Hydraulic Fracturing in Shale Gas Reservoirs*, 2008; http://energy.wilkes.edu/PDFFiles/Issues/EvaluatingTheEnvironmentalImplicationsOfHydraulicFracruringInShaleGasReservoirs.pdf (last accessed 16/01/2014).

32. Water UK Report. *Towards Sustainability 2005–2006*, Water UK, London, 2006.

33. State of California. *California Code of Regulations, Title 14: Natural Resources, Division 2: Department of Conservation*, 2007, 18;: ftp://ftp.consrv.

ca.gov/pub/oil/publications/PRC04_January_11.pdf (last accessed 16/01/
2014).

34. Department of Energy and Climate Change. *Bowland Shale Gas Study*,
    2013; https://www.gov.uk/government/publications/bowland-shale-gas-
    study (last accessed 16/01/2014).

35. IGas Energy, *Unlocking Britain's Energy Potential*; http://www.igasplc.
    com/ (last accessed 16/01/2014).

36. Dart Energy, *Portfolio of Assets*; http://www.dartgas.com/ (last accessed
    16/01/2014).

37. Department of Energy and Climate Change. *UK Government Statistics
    2013. Digest of United Kingdom Energy Statistics (DUKES). Natural Gas*, ch.
    4, 2011; https://www.gov.uk/government/uploads/system/uploads/
    attachment_data/file/65800/DUKES_2013_Chapter_4.pdf (last accessed
    16/01/2014).

38. BBC News, *US Shale Gas to Be Exported Globally in $8bn Deal*, 2011;
    http://www.bbc.co.uk/news/business-15464867?utm_source-
    twitterfeed&utm_medium-twitter (last accessed 16/01/2014).

39. American Petroleum Institute, *Report on Liquefied Natural Gas exports –
    America's Opportunities and Advantages*, 2013; http://www.api.org/policy-
    and-issues/policy-items/lng-exports/liquefied-natural-gas-exports-
    americas-opportunity-and-advantage (last accessed 16/01/2014).

40. P. Bloxham and A. Richardson, Australia's Growing Role in Asian Gas,
    2013; www.morningstar.com.au/funds/article/asian-gas/6179?
    q = printme (last accessed 16/01/2014).

41. HM Government, *The Environmental Permitting (England and Wales)
    Regulations 2010*, 2010; www.legislation.gov.uk/ukdsi/2010/
    9780111491423 (last accessed 16/01/2014).

42. Department for Energy and Climate Change, *Oil and Gas Onshore Ex-
    ploration and Production Publications, Guidance and Data including Map-
    ping, Seismic Activity, Wells, and Licensing and Regulation for Onshore Oil
    and Gas*, 2013; http://og.decc.gov.uk/en/olgs/cms/explorationpro/
    onshore.aspx (last accessed 16/01/2014).

43. Institution of Gas Engineers and Managers, *IGEM/G/101, Onshore Nat-
    ural Gas Extraction. Guidance*, 2012; http://www.igem.org.uk/media/
    291311/igem-g-101.pdf (last accessed 16/01/2014).

44. Polish Geological Institute National Research Institute Report, *Financial
    Times*, 5th June 2013.

45. D. Carrington, *Fracking will not Reduce UK Gas Prices*, 2013; http://www.
    theguardian.com/environment/2013/nov/29/browne-fracking-not-
    reduce-uk-gas-prices-shale-energy-bills (last accessed 16/01/2014).

46. Defra, *National Atmospheric Emissions Inventory, Emission Factors for
    Transport*, 2013; http://naei.defra.gov.uk/data/ef-transport (last accessed
    3rd February 2014).

47. J. Broderick, K. Anderson, R. Wood, P. Gilbert and M. Sharmina, *Shale
    Gas: an Updated Assessment of Environmental and Climate Change impacts,*

2011; http://www.co-operative.coop/Corporate/Fracking/Shale%20gas%20update%20-%20full%20report.pdf (last accessed 16/01/2014).

48. Regeneris Consulting. *Economic Impact of Shale Gas Exploration & Production in Lancashire and the UK, A Report prepared for Cuadrilla Resources*, 2011; http://www.cuadrillaresources.nl/wp-content/uploads/2012/02/Full_Report_Economic_Impact_of_Shale_Gas_14_Sept.pdf (last accessed 16/01/2014).

49. Advanced Resources International, *EIA/ARI World Shale Gas and Shale Oil Resource Assessment. Technically Recoverable Shale oil and Shale Gas Resources: An assessment of 137 Shale Formations in 41 Countries Outside of the United States*, 2013, Table 6; http://www.adv-res.com/pdf/A_EIA_ARI_2013%20World%20Shale%20Gas%20and%20Shale%20Oil%20Resource%20Assessment.pdf (16/01/2014).

50. R. L. Santoro, R. H. Howarth, and A. R. Ingraffea, *Indirect Emissions of Carbon Dioxide from Marcellus Shale Gas Development*, 2011; http://www.eeb.cornell.edu/howarth/IndirectEmissionsofCarbonDioxidefrom MarcellusShaleGasDevelopment_June302011%20.pdf (16/01/2014).

51. Department of Energy and Climate Change. *Bowland Shale Gas Report*, 2013; https://www.gov.uk/government/uploads/system/uploads/attachment_data/file/226874/BGS_DECC_BowlandShaleGasReport_MAIN_REPORT.pdf (last accessed 16/01/2014).

52. U.S. Energy Information Administration Information, *U.S. Natural Gas Wellhead Price*, 2013; http://www.eia.gov/dnav/ng/hist/n9190us3m.htm (16/01/2014).

# Shale Gas Boom, Trade and Environmental Policies: Global Economic and Environmental Analyses in a Multidisciplinary Modeling Framework

FARZAD TAHERIPOUR, WALLACE E. TYNER* AND KEMAL SARICA

## ABSTRACT

While shale oil and gas are controversial in some quarters due to perceived environmental risks, there is little doubt that the shale oil and gas boom is having a major positive impact on the US economy. But how large is it? We use a global general equilibrium economic model to estimate the economic impacts of the shale oil and gas technology on the US economy. We have made simulations with and without expansion in shale resources (positive and negative shocks). From 2008 through 2035 the US GDP on average would be 2.2% higher than its 2007 level with the expansion in shale resources (positive shock). Without the expansion in shale resources (negative shock) on average the US GDP would be 1.3% lower than its 2007 level during the same time period. That means that US GDP over the entire period of 2008–35 on average is projected to be 3.5% higher than it would have been without the shale boom. The welfare impacts are also quite large. On average, the welfare difference between the positive shock and the negative shock is $473 billion per year over the period 2008–35. If gas exports are restricted, the magnitude of the annual gains increases to $487 billion.

---

*Corresponding author.

---

Issues in Environmental Science and Technology, 39
Fracking
Edited by R.E. Hester and R.M. Harrison
© The Royal Society of Chemistry 2015
Published by the Royal Society of Chemistry, www.rsc.org

Other impacts are important as well. The shale boom creates substantial employment growth, with jobs growing on average about 1.8% in the positive shock and declining about 1.1% in the negative shock for a net of about +2.9% employment gains. With the shale expansion, oil and gas prices drop by 6% and 16%, respectively, in the period 2007–35. This price drop stimulates an expanded economic activity. If gas exports are restricted, natural gas prices drop 24.1%, providing additional economic stimulus.

## 1 Introduction

In this chapter we take on the daunting task of estimating the overall economic and environmental impacts of the shale oil and gas boom in the US. We do this using a modified version of a well-known computable general equilibrium (CGE) economic model (GTAP). We use the improved model and database to first project what would have been the economic and environmental impacts had the shale boom not occurred (a negative productivity shock representing reduction in gas and oil resources in the absence of new shale resources). Then we simulate the economic and environmental impacts of the shale boom by positively shocking productivity in oil and gas production. Finally, we examine the important issue of the impacts of liquefied natural gas export restrictions. The chapter is organised as follows: (1) review of relevant literature; (2) description of the standard GTAP energy model; (3) description of the modifications we have made to the GTAP model and database for this analysis; (4) presentation of the three modeling experiments; (5) simulation results; and (6) finally, we conclude with a discussion of the major findings.

## 2 Literature Review and Background

Recently, a limited number of studies have tried to quantify the economic and environmental consequences of the new developments in the US energy market. Brown *et al.*[1] examined how the expected expansion in gas shale will affect the US energy market and the link of this expansion to climate policies. These authors examined six different scenarios using the National Energy Modeling System (an energy-economy market equilibrium model – NEMS) and reached two major conclusions: (1) more abundant gas supplies will expand uses in most economic activities; and (2) the expansion in supply of natural gas in combination with appropriate carbon polices will help the US economy to achieve low-carbon standards in future.

Paltsev *et al.*[2] also analysed the expansion in US natural gas. These authors highlighted the uncertainties in the scale and cost of gas resources, the costs of competing alternative potential technologies and mitigation emissions policies. They used two computable general equilibrium models and concluded that: the expansion in shale gas resources will increase the supply of gas for several decades; the costs of expansion in gas resources are uncertain; and even without additional greenhouse gas (GHG) abatement policies, gas production and use in 2050 will be higher than today. Jacoby et al.[3] also examined the influences of the expansion in US shale gas on US energy and environmental policy and reached similar conclusions.

Recent research in this area also examined the consequences of alternative gas export policies for the US economy. NERA Economic Consulting,[4] based on a set of simulation results obtained from an energy-economy model, suggested that an expansion in exports of natural gas will: (1) increase its domestic prices; (2) generate minor gains for the US economy; (3) improve welfare of owners of natural gas resources; and (4) negatively affect energy-intensive industries and electricity producers. Deloitte,[5] using an integrated industry model (named North American Power, Coal and World Gas), reached the expected conclusion that an increase in US gas exports will increase its domestic gas price and lower world price. Brooks,[6] who used the GPCM® Natural Gas Market Forecasting System modeling framework, also made the same argument. Finally, Ditzel et al.[7] carried out a study related to US natural gas export policies and assessed the impacts of those policies on the US manufacturing sector and natural gas prices. These authors concluded that higher gas prices due to the expansion in exports of natural gas will negatively affect manufacturing industries and electricity producers, limit the use of vehicles which operate using natural gas and, in aggregate, negatively affect the entire US economy.

In addition to these studies, others have tried to measure the economic benefits of the expansion in US shale gas. For example, IHS Global Insight Inc.[8] evaluated the economic impacts of the expansion in supply of gas for the US economy using a US macro model in combination with the IMpact analyses for PLANning (IMPLAN) model and made several projections. One is that shale gas will contribute about $118 billion and $231 billion to the US GDP in 2015 and 2035, respectively. They also projected that the gas industry will provide about 0.9 million and 1.6 million job opportunities in these years. Citi GPS,[9] also using a US macro model, concluded that the expansion in gas supply will improve the US GDP by 2–3% by 2020. Finally, Arora,[10] using a US macro computable general equilibrium model, concluded that the expansion in supply of natural gas will improve the US GDP at only a small-to-moderate rate.

These studies focused on shale gas, but did not explicitly cover shale oil as well. This chapter fills the gap in this area and examines the economic and greenhouse gas (GHG) emission consequences of the expansion in US supplies of oil and gas. It also examines the impacts of alternative trade policies for exports of natural gas.

## 3 The GTAP Model

The modeling framework used consists of a revised version of the Global Trade Analysis Project-Energy (GTAP-E) model and an enhanced version of the Market Allocation-Macro (MARKAL-Macro) model. The GTAP-E model developed at Purdue University[11,12] is a CGE economic top-down model which is designed to assess regional economic and environmental impacts of national and multi-national energy-economy-environmental-trade policies. Alternative versions of this model are used in many economic and environmental analyses.[13-16] The model traces production, consumption and trade of goods and services, categorised in several groups, by region. It considers substitution among energy sources and between capital and energy and takes into account competition for energy and other resources among firms. It also manages allocation of energy between households and firms. We modified and extended this model and its database to correctly analyse expansion in US shale gas and its consequences for other energy sources and economic activities.

To help with the GTAP model and database modification, we used a modified version of the MARKAL-Macro model, which is a bottom-up dynamic, perfect-foresight and energy technology-rich linear programming model widely applied in the energy area. In its standard formulation, its objective function is the minimisation of the discounted total system cost, which is formed by summation of capital, fuel and operating costs for resource, process, infrastructure, conversion and end-use technologies.[17] Sarica and Tyner revised the MARKAL model according to new developments in the energy sectors and, in particular, with respect to recent expansion in US biofuel production and, more importantly, in response to the US shale gas boom.[18]

To develop the GTAP modeling framework and database for this analysis, we began with the GTAP database version 8 which represents the world economy in 2007.[19] We made several major modifications in the GTAP-E modeling framework and database to make it more consistent with independent data sources and the MARKAL model projections for the energy market. Then several experiments were developed to evaluate the regional economic and GHG emission impacts of the US shale gas boom under alternative trade policy configurations.

GTAP is a multi-commodity and multi-regional computable general equilibrium model documented in Hertel.[20] It uses constant returns-to-scale production functions and assumes markets are all competitive. In this model each region is represented by a regional household. As shown in Figure 1, the regional household (*e.g.* the United States) collects all the income (including taxes and payments to the primary factors of production such as labour, land, capital and natural resources) in its region and spends it over three expenditure types – private household (consumer), government and savings. The model divides the supply side of the economy into several distinct sectors. Firms pay wage/rental rates to the regional household in return for the employment of land, labour, capital and natural resources. Firms sell their output to other firms (intermediate inputs), private households, government and investment.

Since GTAP is a global model, firms also export tradable commodities and import intermediate inputs from other regions. The government and private

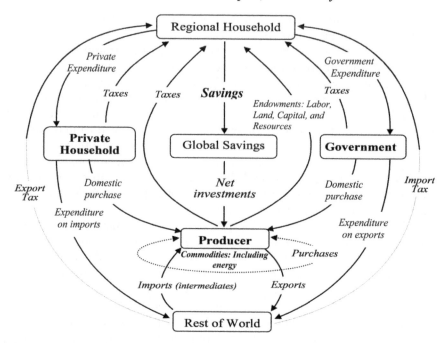

**Figure 1**  An overview of the GTAP model.

household of each region use imported commodities from other regions as well. The imported commodities are assumed to be differentiated by region, and so the model can track bilateral trade flows. Taxes (and subsidies) go as net tax revenues (subsidy expenditures) to the regional household from private household, government and the firms. The rest of the world gets revenues by exporting to private households, firms and government. These revenues are spent on export taxes and import tariffs, which eventually go to the regional household. This rest of world composite is actually made up of many other regions.

On the supply side, the standard GTAP model assumes some degree of substitution among primary factors of production, but it considers no substitution among intermediate inputs or between the composites of intermediate inputs and primary inputs. The GTAP-E model alters this basic set up by introducing substitution among alternative sources of energy used by firms and substitution between the energy bundle and capital.

While the GTAP-E model makes the substitution among energy items possible, its nesting structure does not reflect the recent evolution in energy markets. In recent years a more direct substitution has been observed between gas and coal. Sarica and Tyner demonstrated this fact using the MARKAL-Macro model.[18] The original nesting structure implemented in the GTAP-E model ignores this reality, because it first substitutes oil and gas and then combines the mix of gas and oil with coal. As outlined in the next section, we altered the GTAP-E energy nesting structure to directly substitute gas and coal.

# 4   GTAP Model and Database Modifications

We made four significant changes in the original GTAP-E model and its database. The first changes the representation of natural gas in the GTAP database. The second restructures the GTAP-E intermediate demand structure for energy items and key parameters. The third restructures the depiction of natural resources in GTAP-E to better represent expansion in shale oil and gas. The fourth changes the base assumption on unemployment to reflect the current reality of unemployed labour in the US economy. Each of these is described in turn below.

## 4.1   Natural Gas in the GTAP Database

The database used in this chapter is built on the most up-to-date version of the GTAP database (version 8), which represents the global economy in 2007. The original database divides the whole world into 57 sectors and 129 commodities. To concentrate on important regions and sectors we aggregated the database into 19 regions and 14 sectors. The original GTAP database represents all economic activities related to gas and gas distribution into two separate industries of "gas" and "gdt". The first industry extracts, refines, and processes gas and sells it to some major industries, large-scale electricity producers (big power plants), and the gas distribution network (or "gdt"). The second industry buys gas from the "gas" industry and distributes it among end users: mainly small industries, farms, commercial sector and households. In our aggregation process we kept these two industries separate as they appear in the original database to provide a better picture of the impacts of expansion in shale gas on gas production and consumption. However, following careful examination of the GTAP database version 8, we determined that it suffers from the following major deficiencies in representing monetary values associated with the "gas" and "gdt" sectors:

- The regional monetary values of production and consumption of gas are underestimated.
- The regional monetary values of gas sold by the "gas" industry to the "gdt" industry are ignored.
- The regional distributions of monetary values of gas sold to industries (including power plants) and household do not match independent observations.

In addition to these general issues, we observed an outlier component in the cost structure of the "gdt" sector in US input–output table. To fix these issues we pooled the "gas" and "gdt" together, calculated missing items, re-evaluated the production and consumption of gas, fixed the outlier item in the US input–output table and redistributed monetary values of gas distributed among industries and households using the *GTAPAdjust* program.[21] Finally, we reconstructed the "gas" and "gdt" sectors using the *SplitCom* program.[22]

In the revised database, as we expected, the "gas" industry uses resources and other primary inputs to extract gas, process and refine it, and finally sells produced gas to the main industries, power plants and the gas distribution industry, "gdt", at a wholesale price. The "gdt" sector distributes the gas among its users, which are mainly households, commercial sectors and non-basic industries with a mark-up over the wholesale price.

## 4.2   Firms' Demand for Energy Items

Direct substitution between coal and gas has been observed in firms' demand for energy in recent years. Given the expected expansion in gas resources, this phenomenon is anticipated to prevail in future as well. To make the GTAP-E model consistent with this reality, we introduced a new nesting structure for firms' energy demand. Unlike the original model, at the very bottom nest of the new nesting structure, coal and gas (provided either by "gas" or "gdt") are mixed to represent the direct substitution between gas and coal. Then at a higher level the combination coal–gas is mixed with oil and petroleum products to generate the non-electricity energy input. From this point to the top production function, we preserved the original GTAP-E nesting structure.

For the elasticities of substitution, we rely on the original GTAP values except for the following items. The original GTAP-E model assigns a value of 0.5 to the substitution elasticity between capital and energy. This is a key parameter, which directly affects the changes in the mix of capital and energy in response to the changes in the relative prices of capital and energy. For example, when the value of this elasticity is non-zero, a reduction in the price of energy compared to the price of capital increases the demand for energy, which eventually leads to an increase in the cost share of energy and energy intensity. In a set of trial simulations we learned that compared to the MARKAL model, the GTAP model generates larger changes in the energy intensity due to the changes in the prices of energy items. The MARKAL model uses a value of 0.4 for this parameter, so we adopted this value in the revised GTAP-E model as well.

Finally, we assumed that the elasticity of substitution between coal and gas is 1 to make the demand for coal and gas responsive to changes in their relative prices. We also assumed that the substitution elasticity between the mix of coal–gas and oil is equal to 1. The latter assumption makes the demand system responsive to the changes in relative prices of coal, gas and oil products.

## 4.3   Depiction of Natural Resources

The expected boom in shale oil and gas production in the US is due to the fracking technology, which expands the existing resources. To introduce the expansion in these resources we altered the way that the GTAP-E model handles natural resources. The original GTAP-E model has a pool of resources, a sluggish endowment, and several sectors (including forestry, fishing, mining, coal, oil and gas) share the pool. To model the expansion in oil and gas resources, we divided the pool of resources into two different categories. The forestry, fishing, mining and coal industries use the first group of natural resources, and the gas and oil sectors use the second group. Then we introduced a closure, which expands natural resources used in the oil and gas sectors due to technological progress.

## 4.4    Treatment of Unemployment

The last modification which we introduced into the revised model is related to the supplies of labour and capital. The original GTAP-E model assumes that all factors of production (labour, land, capital and resources) are fully employed. This assumption is not appropriate for today's US economy, which currently suffers from a high rate of unemployment. Under this condition, expansion in shale oil and gas industries could generate major job opportunities, either directly or indirectly. This could also improve and increase investment opportunities. To capture these opportunities and match the model with current conditions, we assume supplies of labour (skilled and unskilled) and capital respond to their prices. That is, rather than assuming full employment, we assume a supply curve for labour and capital. This assumption is introduced into the model with constant elasticity supply functions for labour and capital. We do not alter the full employment assumption for land and resources.

## 5    Simulation Experiments

This chapter aims to evaluate the economic and environmental consequences of the expected expansion in production of shale gas and oil. DOE projections indicate that the US oil and gas production will increase due to the new extraction technologies in the future.[23] As shown in Figure 2, according to these projections, production of oil and gas will increase from 1853 million barrels oil equivalent (MBOE) and 3661 MBOE in 2007 to 2285 MBOE and 5956 MBOE in 2035, respectively. With no expansion in shale resources, supplies of these energy commodities would drop to 1298 MBOE and 3043 MBOE in this year, respectively. Figure 2 shows that with new technologies, supplies of these energy sources will increase rapidly until 2017. After that, the supply of gas will continue to grow, but supply of oil starts to decline slowly and converges to a level higher than its original level in 2007.

Since supplies of oil and gas will follow different expansion paths under alternative scenarios of with and without shale resources, and their expansion/contraction paths vary over time, we divided the time period of 2007 (our benchmark base year) to 2035 (the end year) into five time slices of: 2007–12, 2012–17, 2017–22, 2022–27 and 2027–35. The percentage changes in supplies of oil and gas with and without expansion in shale resources are shown in Figure 2. In essence, we are assuming the DOE forecasts, which come out of an energy-economic model, are the best representation of shale

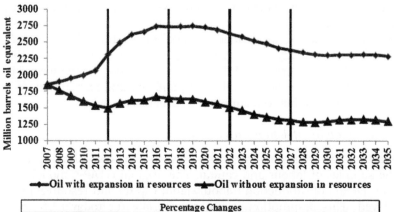

| Percentage Changes | | | | | |
|---|---|---|---|---|---|
| Future trends | 2007-12 | 2012-17 | 2017-22 | 2022-27 | 2027-35 |
| Oil with expansion in resources | 24.8 | 18.4 | -3.7 | -9.8 | -3.9 |
| Oil without expansion in resources | -19.2 | 10.1 | -8.4 | -13.2 | -0.9 |

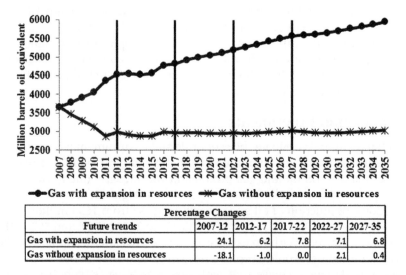

| Percentage Changes | | | | | |
|---|---|---|---|---|---|
| Future trends | 2007-12 | 2012-17 | 2017-22 | 2022-27 | 2027-35 |
| Gas with expansion in resources | 24.1 | 6.2 | 7.8 | 7.1 | 6.8 |
| Gas without expansion in resources | -18.1 | -1.0 | 0.0 | 2.1 | 0.4 |

**Figure 2**   US supplies of oil (upper graph) and gas (lower graph) with and without expansion in shale resources based on the DOE projections.[23]

technology advances. To quantify the impacts of changes in supplies of oil and gas we developed the following two core alternative experiments to simulate impacts of changes in supplies of oil and gas with and without shale resources for each of these five-year time slices:

- *Experiment I*: Changes in US oil and gas with no expansion in shale resources. That is, only conventional oil and gas are included, and therefore total production declines.

- *Experiment II*: Changes in US oil and gas with expansion in shale resources. We also assume no growth in crude oil exports, as exports are prohibited under current US policy.

In these two experiments we assumed that labour and capital are not fully employed and their supplies go up due to higher wages and rental rates. To implement these shocks we assumed that the new fracking technology expands access to more resources at no added cost, which means the new technology shifts out the effective natural gas and crude oil supply curves. In other words, the technology makes more oil and gas available without much of an increase in cost. We defined closures which expand/contract oil and gas resources to achieve the production targets in each time period. In these experiments we make no changes in the tax/subsidy rates, including trade barriers except for crude oil. Current US law does not permit crude oil exports, except for a small amount to Canada. In the second experiment, we fixed the US exports of crude oil due to this law. To isolate the impacts of the new fracking technology, we assumed no exogenous shocks in economic factors, which may affect economic growth at the global scale.

The expansion in shale resources has raised an important debate regarding US trade policy on natural gas exports. Natural gas exports to non-free trade destinations require DOE approval. Some experts argue that the US should allow exports of gas and oil, while others believe that restricting exports of these commodities will be more beneficial for the US economy. In response to this debate we devised another experiment, which restricts natural gas exports in the presence of expansion in shale gas production. The third experiment is:

- *Experiment III*: Changes in US oil and gas with expansion in shale resources, with no change in crude oil or natural gas exports. Petroleum product exports are free to expand.

Finally, in each experiment a five-step process was used to simulate the impacts of changes in oil and gas over time according to the percentage changes in supplies of these energy products as presented in Figure 2. For example, we developed the following five simulations for the first experiment. In the first step a closure was used to change supplies of oil and gas by − 19.2% and − 18.1%, respectively, for the period 2007–12. Then the updated base data obtained from the first step, in combination with a closure, was used to expand supplies of oil and gas by 10.1% and − 1%, respectively, for the period 2012–17. In the third step the up-dated databases of step two in combination with − 8.4% (for oil) and 0.05% (for gas) shocks were implemented for the period 2017–22. In the fourth step the up-dated database of step three in combination with − 13.2% (for oil) and 2.1% (for gas) shocks were implemented for the period 2022–27. In the last step, the up-dated database of step four in combination with − 0.9% (for oil) and 0.4% (for gas) shocks were used for the period 2027–35. This process was repeated in the second and third experiments with the percentage changes corresponding to the case with expansion in shale oil and gas as shown in Figure 2.

## 6  Simulation Results

The expansion in oil and gas directly and indirectly affects the US economy and improves its position in the global energy market. The simulation results obtained from the implemented experiments highlight some important consequences that are outlined in the following analyses.

### 6.1  *Production*

In the absence of expansion in shale oil and gas, US oil and gas production are expected to decrease over time. The simulation results obtained from the first experiment show that in this case almost all sectors, except for the coal industry, will have output falling during the period 2007–35 (see the last column of the top panel of Table 1). However, as shown in Table 1, the rate of reduction varies across sectors over time. Some sectors may experience temporary increases in their output in 2012–17 when US oil production is expected to increase even without new shale resources (see the second numerical column in the first panel of Table 1). In general, in the absence of shale oil and gas, the non-energy sectors (except crops) are expected to lose around 0.5% (for energy-intensive industries) to 1.5% (for services) of their output during the period 2007–35. Reduction in energy sources does not affect crop production in this case because of the weak link between energy production and prices and agriculture.

When expansion of shale oil and gas resources is included in the simulation (*Experiment II*), oil and gas supplies increase by 25.8% and 52%, respectively, during the period 2007–35, and that leads to expansion in production of almost all sectors (except coal) over time. As shown in the last column of the bottom panel of Table 1, the non-energy sectors (except for crops) will experience expansion in their outputs ranging from 0.7% (for energy-intensive industries) to 2.4% (for services) during the period 2007–35.

When exports of gas are limited, expansion in shale oil and gas pushes sectoral outputs to higher levels compared to the case of no trade restriction on gas. This shows that limiting exports of gas results in gains for the US economy in terms of expansion in production.

Among energy sources, the supply of coal behaves differently. Unlike other energy items, the supply of coal increases (by 1.2%) in *Experiment I* and decreases in *Experiments II* (by − 2.1%) and *III* (by − 5.6%) during the period 2007–35. In the first experiment, the expansion in coal production/ consumption happens due to the net of substitution and expansion effects. When supplies of oil and gas decrease, as shown in Table 1, the levels of economic activity and employment turn down, and that reduces demand for all energy items (including coal). This is a negative expansion effect for coal. However, in this case the supply of coal expands in response to shortages in

**Table 1** Changes in production by sector/commodity due to changes in oil and gas resources (figures represent % changes).

| Scenario | Sectors | 2007–12 | 2012–17 | 2017–22 | 2022–27 | 2027–35 | 2007–35 |
|---|---|---|---|---|---|---|---|
| *Experiment I:* With no expansion in shale oil and gas resources | Crops | 0.0 | 0.0 | 0.0 | 0.0 | 0.0 | 0.0 |
| | Livestock | −0.8 | 0.2 | −0.2 | −0.2 | 0.0 | −1.1 |
| | Forestry | −0.6 | 0.1 | −0.1 | −0.1 | 0.0 | −0.8 |
| | Fishing | −0.8 | 0.2 | −0.2 | −0.2 | 0.0 | −1.0 |
| | Food | −0.9 | 0.2 | −0.2 | −0.2 | 0.0 | −1.2 |
| | Coal | 1.1 | 0.0 | 0.1 | 0.0 | 0.0 | 1.2 |
| | Oil | −19.2 | 10.1 | −8.4 | −13.2 | −0.9 | −31.6 |
| | Gas | −18.1 | −1.0 | 0.0 | 2.1 | 0.4 | −16.6 |
| | Gas Distribution | −5.8 | −0.2 | −0.1 | 0.4 | 0.1 | −5.5 |
| | Oil Products | −3.2 | 1.2 | −1.1 | −1.6 | −0.1 | −4.8 |
| | Electricity | −1.3 | 0.1 | −0.1 | 0.0 | 0.0 | −1.4 |
| | Energy Intensive Industries | −0.4 | 0.1 | −0.1 | −0.1 | 0.0 | −0.5 |
| | Other Industries | −0.7 | 0.1 | −0.1 | −0.2 | 0.0 | −0.9 |
| | Services | −1.2 | 0.2 | −0.2 | −0.3 | 0.0 | −1.5 |
| *Experiment II:* With expansion in shale oil and gas resources | Crops | 0.0 | 0.0 | 0.0 | 0.0 | 0.0 | 0.0 |
| | Livestock | 1.0 | 0.6 | 0.1 | −0.1 | 0.1 | 1.7 |
| | Forestry | 0.7 | 0.4 | 0.0 | −0.1 | 0.0 | 1.2 |
| | Fishing | 1.0 | 0.6 | 0.0 | −0.1 | 0.0 | 1.4 |
| | Food | 1.1 | 0.6 | 0.1 | −0.1 | 0.1 | 1.9 |
| | Coal | −1.3 | −0.4 | −0.2 | −0.1 | −0.1 | −2.1 |
| | Oil | 24.8 | 18.4 | −3.7 | −9.8 | −3.9 | 25.8 |
| | Gas | 24.1 | 6.2 | 7.8 | 7.1 | 6.8 | 52.0 |
| | Gas Distribution | 6.5 | 1.5 | 1.5 | 1.2 | 1.1 | 11.8 |
| | Oil Products | 4.1 | 3.3 | −0.5 | −1.7 | −0.5 | 4.8 |
| | Electricity | 1.5 | 0.5 | 0.3 | 0.2 | 0.3 | 2.8 |
| | Energy Intensive Industries | 0.5 | 0.2 | 0.0 | −0.1 | 0.0 | 0.7 |
| | Other Industries | 0.9 | 0.5 | 0.1 | −0.1 | 0.0 | 1.4 |
| | Services | 1.4 | 0.8 | 0.1 | −0.1 | 0.1 | 2.4 |

the energy market, and that leads to an expansion in demand for coal (and eventually in its supply). This is the substitution effect. In the first experiment the net of the expansion and substitution effects is positive, and that increases the production/consumption of coal in the absence of expansion in oil and gas. On the other hand, the expected expansion in the crude oil and gas production (in *Experiments II and III*) improves the level of economic activity and increases employment. That leads to higher demand for energy items, including coal. However, the expansion in oil and gas reduces the prices of these items due to the substitution effect. This eventually leads to a reduction in demand for coal, in particular when exports of gas are limited.

## 6.2   GDP Impacts

From a macro perspective, Figure 3 represents the expected changes in the US GDP at constant prices for the three main experiments over time. This figure indicates that, in the absence of new shale technologies (*i.e.* in *Experiment I*), US GDP drops by 1.6% during the period 2007–35. In this case, the rate of reduction would be slightly smaller in 2007–17 and 2017–22. From 2008 to 2035 on average the US GDP will be 1.3% lower than its original level in 2007. With the expected expansion in shale oil and gas, the US GDP goes up by 2.6% in *Experiment II* during the period 2007–35. From 2008 to 2035 on average the US GDP will be 2.2% higher than its original level in 2007. This average number for *Experiment III* (where we restrict exports of gas) is about 2.3%. In fact, expansion in shale resources on average improves the US GDP by 3.5% (the difference between the average impacts on GDP obtained from *Experiments I and II*). The difference reaches 3.6% if we restrict exports of gas. That is to say, US GDP is expected to be 3.5–3.6% higher due solely to availability of shale technology. This is a level change and not a growth rate.

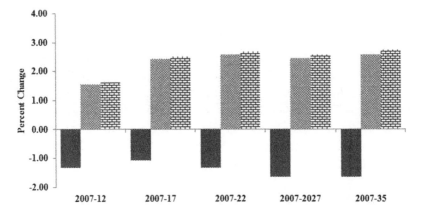

**Figure 3**   Percent changes in US GDP at constant prices.

## 6.3   Employment

The simulation results obtained from the first experiment show that without expansion in crude oil and gas, the demands for unskilled labour, skilled labour and capital will go down by about 1.3%, 1.2% and 1.6%, respectively, during the period 2007–35. With the expansion in shale oil and gas, the demands for these primary inputs increase by 2.1% 1.9% and 2.5%, respectively, in this time period. Our simulation results indicate that adding a restriction on gas exports reshuffles the employment among sectors but will not improve it. These figures confirm that the expansion in shale oil and gas positively affects employment and generates job opportunities directly and indirectly, and that restricting gas exports will not enhance these effects.

## 6.4   Prices

With no expansion in shale oil and gas, the prices of oil and gas are expected to increase by 9.3% and 8.8%, respectively, during the period 2007–35. This increases the prices of gas distribution (by 4.8%), petroleum products (by 3.5%) and electricity (by 0.8%). These increases in energy prices negatively affect economic activities, and that reduces the prices of goods and services during the period 2007–35, as shown the last column of the first panel of Table 2. With the expansion in shale oil and gas, prices follow reverse trends. In this case, prices of oil and gas drop by 5.9% and 16%, respectively, during the same time period, as shown in the last column of the bottom panel of Table 2. This reduces the prices of gas distribution (by 9.1%), petroleum products (by 2.9%) and electricity (1.6%). The lower energy prices generate a boom in the economy, and that leads to higher prices for goods and services. When we limit exports of oil and gas, prices of energy items fall much more. For example, in this case the prices of oil and gas drop by 10.8% and 24.1%, respectively, in the period 2007–35. In conclusion, increases in supplies of oil and gas reduce energy prices. However, lower energy prices increase economic activity, and that leads to slightly higher prices of non-energy commodities due to the positive impacts on households' real incomes.

## 6.5   Trade Impacts

The increases in oil and gas supplies in combination with their general impacts on the growth rate of economic activities have the potential to alter the trade balances of goods and services in different directions. The trade balances will change due to changes in imports and exports and also to changes in world commodity prices. For example, with no expansion in shale oil and gas, the US imports more oil and gas. This increases the demand for these commodities in the global market, which eventually raises prices of these items. The reverse happens in the presence of expansion in shale oil and gas. In what follows we only present the changes in net exports.

**Table 2**  Changes in supply prices due to changes in oil and gas resources (figures represent % changes).

| Scenario | Sectors | 2007 − 12 | 2012 − 17 | 2017 − 22 | 2022 − 27 | 2027 − 35 | 2007 − 35 |
|---|---|---|---|---|---|---|---|
| *Experiment I:* With no expansion in shale oil and gas resources | Crops | − 0.3 | 0.0 | 0.0 | 0.0 | 0.0 | − 0.3 |
| | Livestock | − 0.4 | 0.1 | − 0.1 | − 0.1 | 0.0 | − 0.5 |
| | Forestry | − 0.4 | 0.1 | − 0.1 | − 0.1 | 0.0 | − 0.5 |
| | Fishing | − 0.1 | 0.0 | 0.0 | 0.0 | 0.0 | − 0.1 |
| | Food | − 0.3 | 0.1 | − 0.1 | − 0.1 | 0.0 | − 0.4 |
| | Coal | − 0.2 | 0.1 | − 0.1 | − 0.1 | 0.0 | − 0.3 |
| | Oil | 6.4 | − 2.7 | 2.6 | 4.0 | − 0.9 | 9.3 |
| | Gas | 8.9 | 0.4 | 0.0 | − 0.9 | 0.4 | 8.8 |
| | Gas Distribution | 5.1 | 0.3 | − 0.1 | − 0.6 | 0.1 | 4.8 |
| | Oil Products | 2.4 | − 0.9 | 0.9 | 1.2 | − 0.1 | 3.5 |
| | Electricity | 0.9 | 0.1 | 0.0 | − 0.1 | 0.0 | 0.8 |
| | Energy Intensive Industries | 0.0 | 0.0 | 0.0 | 0.0 | 0.0 | 0.0 |
| | Other Industries | − 0.3 | 0.1 | − 0.1 | − 0.1 | 0.0 | − 0.3 |
| | Services | − 0.3 | 0.1 | − 0.1 | − 0.1 | 0.0 | − 0.4 |
| *Experiment II:* With expansion in shale oil and gas resources | Crops | 0.4 | 0.2 | 0.1 | 0.0 | 0.1 | 0.7 |
| | Livestock | 0.5 | 0.3 | 0.1 | 0.0 | 0.1 | 0.8 |
| | Forestry | 0.5 | 0.3 | 0.0 | 0.0 | 0.0 | 0.8 |
| | Fishing | 0.2 | 0.0 | 0.1 | 0.1 | 0.1 | 0.5 |
| | Food | 0.3 | 0.2 | 0.0 | 0.0 | 0.0 | 0.6 |
| | Coal | 0.3 | 0.2 | 0.0 | − 0.1 | 0.0 | 0.4 |
| | Oil | − 6.6 | − 5.4 | 1.3 | 3.5 | 1.3 | − 5.9 |
| | Gas | − 8.5 | − 1.9 | − 2.2 | − 1.8 | − 1.6 | − 16.0 |
| | Gas Distribution | − 4.9 | − 1.0 | − 1.3 | − 1.0 | − 0.9 | − 9.1 |
| | Oil Products | − 2.9 | − 2.4 | 0.5 | 1.4 | 0.5 | − 2.9 |
| | Electricity | − 0.9 | − 0.1 | − 0.2 | − 0.2 | − 0.1 | − 1.6 |
| | Energy Intensive Industries | 0.0 | 0.0 | 0.0 | 0.0 | 0.0 | 0.1 |
| | Other Industries | 0.3 | 0.2 | 0.0 | 0.0 | 0.0 | 0.5 |
| | Services | 0.4 | 0.2 | 0.0 | 0.0 | 0.0 | 0.7 |

In the absence of expansion in shale oil and gas, the US energy trade balance goes down by about $44 billion due to increases in net imports of energy items in 2007–35 (see the last column of the top panel of Table 3). On the other hand, since the reduction in supply of energy negatively affects employment and that reduces household incomes and their demand for goods and services, the US economy exports/imports more/less non-energy items (by $76 billion net), as listed in the last column of the top panel of Table 3. Hence, in the first experiment the US overall net trade improves by $32 billion in the period 2007–35.

In the second experiment, with the expansion in shale oil and gas, the US net trade in energy items improves, significantly. In this case, the US net export of energy (the combination of reductions in imports and expansions in exports) increases by about $72 billion in the period 2007–35, as shown in the last column of the middle panel of Table 3. However, in this case, due to the expansion in economic activities, increases in employment and improvements in household incomes, demand for goods and services goes up. This reduces the net exports of non-energy commodities by $122 billion in 2007–35. Hence, in the second experiment the US net trade balance worsens by about $50 billion in 2007–35. Finally, in the third experiment, where we restrict exports of both oil and gas, the net export of energy drops to $49 billion in 2035, and the overall net trade balance decreases by about $56 billion as shown in the last column of the bottom panel of Table 3.

## 6.6    Welfare Impacts

Economists use changes in economic well-being (called "welfare") to measure the impacts of economic changes. The expansion in shale oil and gas affects many aspects of the US economy and those of other regions. These changes eventually affect consumer welfare. The GTAP-E model measures changes in welfare using the concept of *Equivalent Variation* (EV), which measures changes in welfare in monetary terms. The regional changes in welfare for the three main experiments are reported in Table 4 for each of the time slices of 2007–12, 2012–17, 2017–22, 2022–27, 2027–35 and for the entire period 2007–35.

This table shows that, with no expansion in shale oil and gas, US welfare goes down by $223 billion in 2035 compared to its 2007 level. In the second experiment, the expansion in shale oil and gas improves US welfare by $354 billion in 2035 compared to its 2007 level. Finally, in the third experiment, where we restrict exports of oil and gas jointly, welfare goes up further to $377 billion in 2035 compared to its 2007 level.

Since the magnitudes of changes in US oil and gas production vary over time, we now provide a more comprehensive analysis of the welfare impacts of the expansion in shale oil and gas based on the results obtained for all five periods 2007–12, 2102–17, 2017–22, 2022–27 and 2027–35. As shown in Figure 4, with no expansion in shale resources US welfare drops during the period 2007–12. The magnitude of welfare losses would be

**Table 3** Changes in commodities trade balances due to changes in shale oil and gas production (1000 US 2007 dollars).

| Scenario | Sectors | 2007-12 | 2012-17 | 2017-22 | 2022-27 | 2027-35 | 2007-35 |
|---|---|---|---|---|---|---|---|
| *Experiment I*: With no expansion in shale oil and gas resources | Agriculture Products and Food | 2952 | -541 | 561 | 687 | 22 | 3680 |
| | All energy items | -34 078 | 7018 | -7134 | -9072 | -337 | -43 602 |
| | Coal | 108 | -34 | 33 | 46 | 2 | 155 |
| | Oil | -15 867 | 7523 | -6825 | -10 378 | -648 | -26 195 |
| | Gas and Gas Distribution | -16 277 | -1389 | 492 | 2483 | 383 | -14 307 |
| | Oil Products | -2037 | 932 | -845 | -1244 | -76 | -3270 |
| | Electricity | -6 | -14 | 12 | 21 | 2 | 14 |
| | Industry and services | 56 943 | -11 035 | 11 330 | 14 062 | 477 | 71 777 |
| | Total | 25 818 | -4558 | 4758 | 5677 | 161 | 31 855 |
| *Experiment II*: With expansion in shale oil and gas resources | Agriculture Products and Food | -3533 | -2022 | -361 | 232 | -337 | -6021 |
| | All energy items | 41 107 | 24 706 | 4398 | -2631 | 4559 | 72 138 |
| | Coal | -136 | -102 | -1 | 30 | -4 | -212 |
| | Oil | 18 429 | 16 525 | -4647 | -11 091 | -4558 | 14 658 |
| | Gas and Gas Distribution | 20 183 | 5832 | 9799 | 10 139 | 9859 | 55 812 |
| | Oil Products | 2633 | 2476 | -761 | -1724 | -740 | 1884 |
| | Electricity | -2 | -26 | 7 | 15 | 2 | -4 |
| | Industry and services | -68 254 | -40 264 | -6444 | 5368 | -6237 | -115 831 |
| | Total | -30 681 | -17 580 | -2407 | 2969 | -2014 | -49 713 |
| *Experiment III*: With expansion in shale oil and gas resources and restriction on gas exports | Agriculture Products and Food | -3482 | -1981 | -263 | 376 | -154 | -5505 |
| | All energy items | 37 127 | 22 565 | 141 | -8325 | -2599 | 48 908 |
| | Coal | -107 | -88 | 26 | 65 | 38 | -67 |
| | Oil | 18 380 | 16 451 | -4756 | -11 244 | -4789 | 14 042 |
| | Gas and Gas Distribution | 15 716 | 3479 | 5123 | 3910 | 2066 | 30 294 |
| | Oil Products | 3083 | 2720 | -317 | -1152 | -19 | 4314 |
| | Electricity | 55 | 2 | 66 | 96 | 105 | 325 |
| | Industry and services | -65 692 | -38 764 | -3287 | 9691 | -886 | -98 939 |
| | Total | -32 048 | -18 180 | -3409 | 1741 | -3640 | -55 536 |

**Table 4** Regional welfare impacts due to changes in shale oil and gas production (1000 US 2007 dollars).

| Experiments | Regions | 2007-12 | 2012-17 | 2017-22 | 2022-27 | 2027-35 | 2007-35 |
|---|---|---|---|---|---|---|---|
| *Experiment I:* With no expansion in shale oil and gas resources | USA | -178 403 | 32 931 | -34 034 | -41 815 | -1339 | -222 660 |
| | Canada | 2628 | 10 | 92 | -89 | -37 | 2603 |
| | Central and South America | 2634 | -520 | 534 | 674 | 23 | 3345 |
| | EU 27 | -3738 | 1819 | -1647 | -2502 | -157 | -6225 |
| | China, India, Japan, & E. & S.E. Asia | -3408 | 1719 | -1550 | -2368 | -150 | -5758 |
| | Russia and other members of FSU | 2793 | -776 | 751 | 1038 | 51 | 3856 |
| | Middle East and North Africa | 5652 | -2125 | 1975 | 2899 | 167 | 8568 |
| | Others | 2876 | -964 | 908 | 1305 | 71 | 4196 |
| | World | -168 967 | 32 093 | -32 972 | -40 859 | -1370 | -212 075 |
| *Experiment II:* With expansion in shale oil and gas resources | USA | 213 079 | 122 195 | 19 229 | -17 535 | 17 145 | 354 113 |
| | Canada | -2680 | -686 | -508 | -305 | -318 | -4497 |
| | Central and South America | -2756 | -1550 | 3 | 550 | 83 | -3670 |
| | EU 27 | 4547 | 4211 | -956 | -2484 | -788 | 4529 |
| | China, India, Japan, & E. & S.E. Asia | 4006 | 3828 | -1108 | -2570 | -999 | 3157 |
| | Russia and other members of FSU | -3559 | -2446 | -473 | 122 | -685 | -7041 |
| | Middle East and North Africa | -6633 | -5232 | 720 | 2580 | 675 | -7890 |
| | Others | -3417 | -2536 | 137 | 972 | 81 | -4763 |
| | World | 202 588 | 117 785 | 17 043 | -18 671 | 15 193 | 333 938 |
| *Experiment III:* With expansion in shale oil and gas resources and restriction on gas exports | USA | 220 171 | 124 730 | 23 244 | -13 142 | 22 026 | 377 031 |
| | Canada | -3458 | -900 | -598 | -120 | 27 | -5049 |
| | Central and South America | -2462 | -1407 | 268 | 851 | 418 | -2331 |
| | EU 27 | 4190 | 4023 | -1280 | -2885 | -1306 | 2742 |
| | China, India, Japan, & E. & S.E. Asia | 4091 | 3831 | -1093 | -2544 | -980 | 3306 |
| | Russia and other members of FSU | -2978 | -2148 | 73 | 826 | 220 | -4007 |
| | Middle East and North Africa | -6279 | -5049 | 1032 | 2942 | 1104 | -6251 |
| | Others | -3139 | -2400 | 378 | 1264 | 431 | -3467 |
| | World | 210 137 | 120 680 | 22 024 | -12 808 | 21 940 | 361 974 |

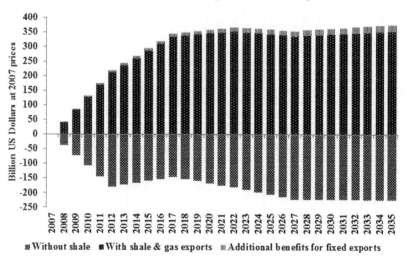

     ⊠ Without shale    ■ With shale & gas exports    ⦀ Additional benefits for fixed exports

**Figure 4**   Annual welfare gains due to expansion in shale oil and gas.

about − \$178 billion by 2012. Then, since US conventional oil production is expected to increase during 2012–17, the magnitude of welfare losses drops slightly by 2017. After that, the magnitude of welfare losses grows until 2027 and then remains at around − \$222 per year until 2035. In general, with no expansion in shale resources, the overall cumulative welfare losses during the time period of 2007–35 would be about \$4957 billion.

On the other hand, with expansion in shale resources, US welfare increases rapidly during the first two time slices of 2007–12 and 2012–17. The magnitude of welfare gains reaches \$355 billion in 2017 and remains at around this figure until 2035. In this case, the overall cumulative welfare gains during 2007–35 would be about \$8298 billion. Restricting exports of gas could add \$372 billion to these cumulative gains. Hence, in general, the expansion in shale resources improves US welfare by \$13 256 billion (the difference between overall welfare losses with no expansion in shale resources and overall welfare gains with shale resources). This overall benefit rises to \$13 628 billion when we impose restrictions on exports of oil and gas. Recall that the US annual GDP is around \$15 trillion. These simulated figures indicate that the shale oil and gas industries are expected to make significant contributions to the future US economy.

The expansion in shale oil and gas affects the welfare of other countries as well. As shown in Table 4, some regions will lose welfare and others will gain. For example, Russia and other member countries of the former Soviet Union, the Middle East and North Africa, Canada and South and Central America, all of which export oil and gas, will be the main losers due to the expansion in energy production in the US. The European Union and many Asian countries will be big winners. These regions gain because of the reduction in global energy prices due to the expansion in US energy. They also gain due to their trade of non-energy items with the US.

## 6.7 Emissions

The new oil and gas extraction technologies expand the supplies of oil and gas, and that lowers the production and consumption of coal. However, the increases in supplies of oil and gas exceed the reduction in coal production, and consequently the overall supply of energy increases. This, in turn, increases economic activity, which results in more emissions. The reverse happens with no expansion in shale oil and gas. In the first experiment the overall level of $CO_2$ emissions generated by the US economy decreases by 2.4% due to reductions in supplies of oil and gas during the period 2007–35. On the other hand, in the second experiment, the overall level of $CO_2$ emissions produced by the US economy increases by 4.1% due to the expansion in supply of energy and economic activity. With restrictions on exports of gas, emissions expand by 6.9% during the period 2007–35. As shown in Figure 5, in the first experiment the index of $CO_2$ emissions per dollar of produced commodities (defined as $CO_2$ emissions over GDP at constant 2007 prices) decreases from 396 g in 2007 to 393 g in 2035. With increases in supplies of oil and gas, this index increases following a steepening trend and reaches to 402 g in 2035 in the second experiment. The lower prices of energy in the second experiment contribute to a higher level of emissions per dollar. In this case, energy intensity goes up in many sectors due to substitution between energy and capital and also substitution among capital–energy and other primary inputs. This leads to a higher level of emissions per unit of output. Banning exports of gas causes the index of $CO_2$ emissions per dollar of GDP to increase to 412 g. This means that if we limit exports of gas then the energy intensity would increase slightly in future, everything else being unchanged.

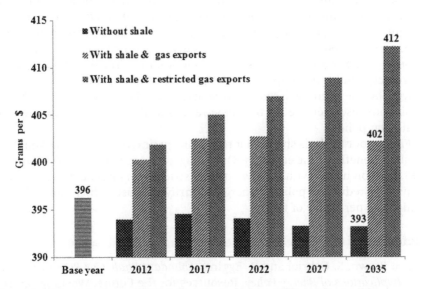

**Figure 5**  $CO_2$ emissions per US dollar production at 2007 constant prices.

In this chapter we have used a modeling framework which assumes perfect competition in all markets, including energy. The implemented model also allows substitution among capital and other primary inputs. In this modeling framework, when energy prices fall, due to increases in their supplies, energy intensity and emissions per unit of output go up. In the real world, however, energy prices may not go down due to imperfect competition in energy markets. For example, OPEC may react to the expansion in US supply of energy and adjust its oil supply. In this case energy prices may not fall as much, and hence emissions per unit of output may not go up. Furthermore, we have not imposed any emissions regulation policy in our model. In practice, the government probably would impose regulations to encourage producers to reduce emissions per unit of output.

## 7 Conclusions

Clearly, the shale oil and gas boom has a major impact on the US economy. Based on the predictions of our model, from 2008 to 2035 the US GDP on average would be 2.2% higher than its 2007 level with the expansion in shale resources. Without that expansion, on average the US GDP would be 1.3% lower than its 2007 level during the same time period. That means that US GDP over the entire period of 2008–35 on average is 3.5% higher than it would have been without the shale boom. The welfare impacts are also quite large. On average, the welfare difference between the positive shock and the negative shock is $473 billion per year over the period 2008–35. If we restrict gas exports, the magnitude of the annual gains increases to $487 billion.

Other impacts are important as well. The shale boom creates substantial employment growth, with jobs growing on average about 1.8% in the positive shock and declining about 1.1% in the negative shock for a net of about + 2.9% employment gains. With shale expansion, oil and gas prices drop by 6% and 16%, respectively, in the period 2007–35. This price drop stimulates expanded economic activity. If gas exports are restricted, natural gas prices drop 24.1%, providing additional economic stimulus. The overall trade deficit increases, unlike the improvement in energy trade balance, interestingly, driven by the increased economic activity stimulated by the shale boom. Emissions also increase for the same reason in the absence of an emission reduction policy.

Finally, our analysis shows that restricting gas exports provides a small but positive benefit for the economy. On average, the annual economic welfare is $13.3 billion higher with the gas export restriction in place. In the absence of emissions reduction policies, energy intensity increases which causes higher emissions per dollar of GDP.

## References

1. S. Brown, S. Gabriel and R. Egging, *Abundant Shale Gas Resources: Some Implications of Energy Policy*, Resources for the Future, Washington, DC, 2010.

2. S. Paltsev, D. Jacoby, J. M. Reilly, Q. Ejaz, F. O'Sullivan, J. Morris, S. Rausch, N. Winchester and O. Kragha, *Energy J.*, 2011, **39**, 5309–5321.

3. D. Jacoby, F. O'Sullivan and S. Paltsive, *The Influence of Shale Gas on U.S. Energy and Environmental Policy*, MIT Joint Program on the Science and Policy of Global Change, Cambridge, MA, 2011.

4. NERA Economic Consulting, *Macroeconomic Impacts of LNG Exports from the United States*, Washington, DC, 2012.

5. Deloitte, *Made in America: The Economic Impact of LNG Exports from the United States*, Center for Energy Solutions, Washington, DC, USA, 2011.

6. R. Brooks, *Using GPCM to Model LNG Exports from the US Gulf Coast*, RBAK Inc., Houston, TX, USA, 2012.

7. K. Ditzel, J. Plewes and B. Broxson, *US Manufacturing and LNG Exports: Economic Contributions to the US Economy and Impacts on US Natural Gas Prices*, Charles River Associates, Washington, DC, 2013.

8. IHS Global Insight, *The Economic and Employment Contributions of Shale Gas in the United states*, IHS Global Insights (USA) Inc., Washington, DC, 2011.

9. Citi GPS, *Energy 2020: North America, The New Middle East?*, Citi GPS: Global Perspecitves and Solutions, New York, NY, USA, 2012.

10. V. Arora, *Energy Information Administration*, U.S. Department of Energy, Washington, DC, 2013.

11. J. Burniaux and T. Truong, *GTAP-E: An Energy-Environmental Version of the GTAP Model*, Purdue University, West Lafayette, IN, 2002.

12. R. McDougall and A. Golub, *GTAP-E Release 6: A Revised Energy-Environmental Version of the GTAP Model*, Purdue University, West Lafayette, IN, 2007.

13. A. Golub, B. Henderson, H. Thomas, P. Gerber, S. Rose and B. Sohngen, *Proc. Natl. Acad. Sci. U. S. A.*, 2012; doi: 10.1073.

14. T. Hertel, A. Golub, A. Jones, M. O'Hare, R. Plevin and D. Kammen, *Bioscience*, 2010, **60**, 223–231.

15. C. Kemfert, M. Kohlhaas, T. Truong and A. Protsenko, *Climate Policy*, 2006, **6**, 441–455.

16. F. Taheripour, T. W. Hertel, W. E. Tyner, J. Beckman and D. K. Birur, *Biomass and Bioenergy*, 2010, **34**, 278–289.

17. R. Loulou, G. Goldstein and K. Noble, *Documentation for the MARKAL Family of Models*, Energy Technology Systems Analysis Program, Paris, France, 2004.

18. K. Sarica and W. E. Tyner, *Energy Econ.*, 2013, **40**, 40–50.

19. B. G. Narayanan, A. Aguiar and R. McDougall, *Global Trade, Assistance, and Production: The GTAP 8 Data Base*, Center for Global Trade Analysis, Purdue University, West Lafayette, IN, 2012.

20. T. W. Hertel, *Global Trade Analysis: Modeling and Applications*, Cambridge University Press, New York, 1997.

21. M. Horridge, GTAPAdjust: A program to balance or adjust a GTAP database, Centre of Policy Studies, Monash University, Melbourne, Australia, 2011. The program and its manual are available at: http://www.copsmodels.com/archivep.htm.

22. M. Horridge, SplitCom - Programs to Disaggregate a GTAP Sector, Centre of Policy Studies, Monash University, Melbourne, Australia, 2005. The program and its manual are available at: http://www.copsmodels.com/splitcom.htm.

23. U.S. Department of Energy, *Annual Energy Outlook*, Washington, DC, 2013. Available for download at http://www.eia.gov/forecasts/archive/aeo13/.

# Exploration for Unconventional Hydrocarbons: Shale Gas and Shale Oil

IAIN C. SCOTCHMAN

ABSTRACT

Over the last 25 years the development of unconventional hydrocarbons, shale gas and shale oil, has come to dominate the oil industry,
particularly in the USA where their development has dramatically
overturned the decline in domestic production, with self-sufficiency in
gas production attained. Exploration for and production of shale-based
resources requires a very different approach and mind-set from that
required for conventional resource exploration and production. The
aim of this chapter is to discuss the geology of shale resources and the
techniques developed for their exploration and exploitation.

## 1 Introduction

Unconventional hydrocarbons comprise oil and gas resources trapped in
any "non-conventional" reservoir (*i.e.* not the porous and permeable sandstones and limestones which comprise "conventional" reservoirs), ranging
from gas hydrates and oil shales located beneath the sea-bed or ground
surface, respectively, through shale gas and shale oil to coal-bed methane
(CBM) (see Figure 1). However, in general, the term "unconventional
hydrocarbons" has come to relate to shale resources and CBM, as gas hy-
drate development is very much in its infancy and oil shale, which cradled
the 19th century UK petroleum industry in central Scotland, requires
considerable energy input to retort the shales, with consequent large

Issues in Environmental Science and Technology, 39
Fracking
Edited by R.E. Hester and R.M. Harrison
© The Royal Society of Chemistry 2015
Published by the Royal Society of Chemistry, www.rsc.org

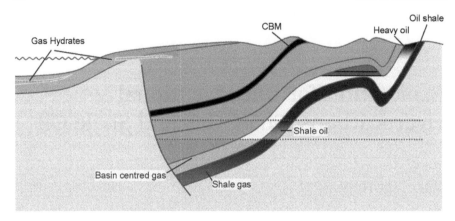

**Figure 1**   Unconventional hydrocarbon systems.

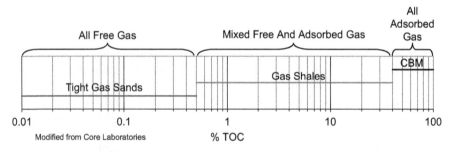

**Figure 2**   Definition of tight gas, shale gas and coal bed methane based on organic
matter content (modified from Core Laboratories).

environmental footprint. Both shale gas and CBM are characterised by very
low matrix permeability and natural fractures, the former generally
comprising organic-rich shales containing both free and adsorbed gas,[1]
while CBM resources are limited to coal seams and only contain adsorbed
gas (see Figure 2). Both shale resources and CBM require the creation of
an artificial reservoir using hydraulic fracturing, colloquially known as
"frac'ing" or "fracking", of horizontal well bores drilled within the near-
impermeable shale or coal. As this chapter relates only to shales, CBM will
not be discussed further.

   Shale-based resources have a long history of exploitation.[2] The first recorded gas production was at Fredonia, Pennsylvania, USA, in 1821 where
Devonian-aged shale provided gas for lighting from a simple well.[3] Similarly,
in the late 19th century, shale-sourced gas at Heathfield, Sussex, lit the
railway station.[4] Gas-charged shales were later recognised by early oil-well
drillers in the USA as a potential hazard, often with fatal explosions and
fires. These shales, generally of Palaeozoic age, were often encountered
during drilling for conventional oils and gas reservoirs across the USA, from

Pennsylvania and Ohio south-westwards to Texas, requiring the early wells to wait several days for these formations to de-gas. Later wells have required expensive steel casing to contain the generally overpressured gas. It was not until the 1980s, with indigenous conventional-reservoired gas supplies declining, that in the USA these were considered as a potential resource.[5,6] Early exploitation of gas from these shales was limited generally to formations containing natural fractures, usually with short-lived production, a notable exception being the Ohio Shale Big Sandy Field in Kentucky, USA, which has been on production since the 1920s.[7] Antrim Shale production followed in the 1930s from the Michigan Basin, but it was the advent of hydraulic fracturing in the late 1940s which aided this development, although not until the mid-1980s did entrepreneurs such as the late George Mitchell actively target shales as a source of gas. Large-scale hydraulic "fracs" were developed but the sought-after breakthrough occurred when they were used sequentially in horizontal well-bores drilled within the gas-rich shale intervals.[8] This has led to the current boom in shale gas and oil production in the USA (see Figure 3). However, whether this success will be replicated in Europe and elsewhere in the world is yet to be seen, with the ever-present attendant environmental concerns generally foremost to any reasoned debate.

**Figure 3**   Map of major US shale plays.
(Source: U.S. Energy Information Administration).[52]

## 2   Shale Resource Systems – Concepts

- Shale plays exhibit high levels of heterogeneity and present complex geology/mineralogy, with "nanopores" and very low matrix permeability, generally with natural fractures, and gas storage related to both porosity and adsorption on both organic matter and clay minerals.
- Comprises a hydrocarbon source rock which is likely to have previously charged "conventional" plays within the basin with oil and gas.
- Non-expelled hydrocarbons retained in source rock – shale gas, although often "liquids-rich", dependent on thermal maturity.
- Local migration into associated more-permeable lithologies – shale oil "hybrid" reservoir.
- No "natural" permeable reservoir: equivalent to trap cap-rock or seal in "conventional" plays – porous but with very small pore throats.
- Reservoir man-made – hence the need for long horizontal wells and hydraulic fracturing.

Shale "plays" can be described as fine-grained reservoirs defined by a microstructure with a $<5$ μm grain size, of variable lithology and mineralogical composition, characterised by nanometer-scale pores ("nanopores") with extremely small pore throats and negligible permeability, which are producible only by multiple-frac'ed horizontal wells. Many, but not all, such reservoirs are organic-rich and are generally hydrocarbon source rocks with a variable thermal maturation state, examples being the Barnett, Marcellus, Haynesville and Eagle Ford Shales. The Niobrara Formation is a typical carbonate-rich, non-source shale reservoir. Mineralogy is variable and is critical to a successful shale play, generally requiring a large proportion of "brittle" minerals such as silica or carbonate forming detrital matrix and cement components compared to "ductile" clay and organic matter components. Silica-rich shale plays include the Barnett, Marcellus, Mowry and Duvernay Shales, while the Eagle Ford, Niobrara and Vaca Muerta formations are carbonate-rich shale plays. The Haynesville Shale is a mixed silica–carbonate shale play but, importantly, none of these shales is totally clay-rich, although the Marcellus can reach up to 50% clay. Natural fractures are commonplace, generally related to local faulting and to development of regional tectonic stress patterns: these can be either open or cemented by minerals such as calcite and can influence the effectiveness of well stimulation. The aim of hydraulic fracturing a well is to achieve a large artificial "flow-unit" or stimulated reservoir volume by creating a complex fracture network to connect the hydrocarbon-bearing "nanopores" with the well-bore. Small-scale local fracture systems can be exploited and extended from the well-bore into the reservoir by the stimulation, but cemented fractures may be deleterious as the cements generally have a lower fracture threshold than the surrounding shale and much of the stimulation energy is diverted into opening these fractures rather than the stronger shale matrix. Gas storage within organic-rich shales occurs as both free gas within the pores of the rock matrix and any fracture system and, importantly, adsorbed onto organic matter and clay mineral surfaces, particularly the walls of organic-matter hosted pores. While free gas predominates,[9] the relative importance of adsorbed gas is highly variable, both within and between shales, varying as a function of the amount and type of organic matter present, pore size distribution, mineralogy, level of diagenesis or maturation state and reservoir conditions (temperature and pressure).[10]

Shale resources represent a complete petroleum system,[11] comprising the hydrocarbon source, reservoir and seal within a single rock unit.[9] They are generally described as "statistical plays", as each well not only accesses a separate portion of the continuous shale reservoir with its very low inherent permeability and variations in geological parameters, but also due to the requirement for the completion process to create an artificially permeable reservoir. Hydraulic fracturing of horizontal reservoir sections, optimally located within the shale, is the key to successful access to such resources. While shale resources can be determined by well log data calibrated to cores, as discussed below, due to the need to artificially create a permeable reservoir post-drilling, conversion to reserves is necessarily made by the use of production data to determine the Estimated Ultimate Recovery (EUR).[6] The EUR for a shale well is largely dependent on the following factors discussed below:

- Shale quality (thickness, richness, reservoir potential);
- Horizontal well-bore length;
- Number of hydraulic fractures created in the reservoir section; and
- Effectiveness of these fractures in accessing the hydrocarbon-bearing pores.

The EUR is calculated from the well initial production rate (IP) in combination with an estimated type decline curve (see Figure 4) and is generally in the range of 15–35% of the calculated gas resource (gas in place).[8]

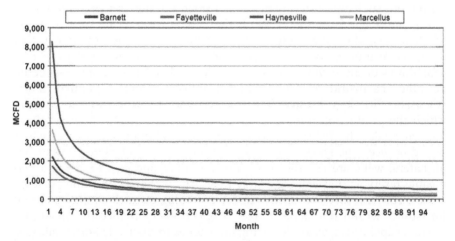

**Figure 4** Typical production decline curves for US gas shales.

## 2.1   Shale Gas

Shale gas resource plays are distinguished by gas type and "system" characteristics,[9] typically with either biogenic or thermogenic gas production[12] from low porosity and permeability shales, forming a self-contained source – reservoir – seal system.

Biogenic gas shales are thermally immature (vitrinite reflectance (VR) <0.5% Ro) due to only shallow burial, typically with formation temperatures of 40–80 °C. At temperatures above 80 °C most bacteria are destroyed by "pasteurisation" and thermally generated gas becomes predominant. Biogenic gas shales include the Antrim Shale Formation in the Michigan Basin,[13] and the shallow buried sections of the New Albany Shale, typified by dry gas adsorbed onto organic matter with, after de-watering, modest initial production (IP) rates of 40–500 million cubic feet of gas per day (mmcf d$^{-1}$) with long production of over 30 years. Thermogenic shale gas plays range in maturity from early gas-window such as the New Albany Shale to the very prolific late gas-window plays such as the Barnett, Marcellus and Haynesville Shales, with EURs for the Barnett Shale in excess of 2.5–3.5 billion cubic feet (BCF)[14] and 5–7 BCF in the high-maturity Marcellus Shale in northern Pennsylvania.

Most shale-gas plays have been affected by tectonic uplift, which effectively shuts off the hydrocarbon generation process by lifting them out of the gas generation "window". Due to the impermeable character of the shales, most of the gas is retained in the rock unless released by fracturing or faulting related to the tectonic activity. This has two benefits: the gas is generally overpressured in uplifted reservoirs and the cost of drilling to reach the shales is reduced.

## 2.2   Shale (Tight) Oil

Shale or 'tight' oil is a more recent concept, with large potential in the USA in formations such as the Devonian Bakken Shale and the Cretaceous Niobrara, Mowry and Eagle Ford Shales. While the basic principles are similar to that of shale gas, the much larger oil molecule size relative to pore throat diameter results in large capillarity effects which greatly restrict the flow capabilities of these shales. To be effective, shale oil plays require what are described as hybrid reservoirs which comprise a coarser grained, more-permeable lithology intimately associated with the shale source rock in which the production wells can drilled and completed. In the case of the Bakken Shale, the hybrid reservoir comprises a limestone unit which is sealed between the Lower and Upper Bakken source rock shales (see Figure 5). Lateral wells are completed within this limestone unit to create a permeable reservoir that can effectively flow the oil charge from the adjacent shales. Thin interbeds of more permeable lithologies, such as siltstones and limestones within the shale source rock, can also enhance the shale production capabilities, examples being limestones in the Eagle Ford Shale (see Figure 6) and siltstones in the Mowry Shale. In these cases, oil production is

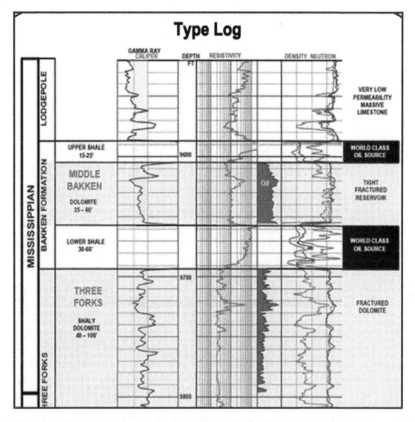

**Figure 5**   Bakken Shale Formation stratigraphic column, illustrating the 'hybrid' shale oil reservoir concept.

achieved by drilling and completing lateral wells in sections where the interbedded units are best developed.

## 3   What Makes a Shale Reservoir?

Several key geological criteria have been developed from the experience gained from exploration for shale plays and their subsequent exploitation:

- *Organic-richness*: a minimum present day total organic carbon (TOC) of 2–3 weight % (wt%) for shale-gas and 4–5 wt% for shale-oil, comprising oil-prone kerogen;
- *Thermal maturity*: an obvious criterion, "gas window" for shale-gas (VR >1.3% Ro) and "oil window" for shale-oil (VR of 0.7–1.3% Ro); and
- *Reservoir*: in shale reservoirs, the key parameters controlling hydrocarbon storage and flow, porosity, permeability and adsorption, are strongly related to the lithology and mineralogy of the rock, to the degree of natural fracturing of the rock and the ability to artificially fracture the rock, which is also affected by the present day tectonic stress regime.

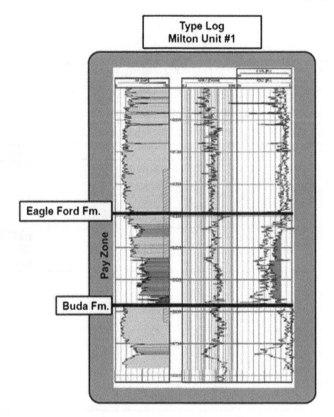

**Figure 6**   Eagle Shale Formation stratigraphic column, illustrating shale oil reservoir.

## 3.1   Organic Richness

The basic prerequisites for a productive thermogenic shale-gas play are an organic-rich shale which has attained a gas-window level of thermal maturity prior to any subsequent uplift. In detail, an effective hydrocarbon source rock is required with development of porosity to store generated but non-expelled hydrocarbons allied with a 'tight', very low permeability matrix to isolate these pores and retain the trapped hydrocarbons.[9] These shales therefore have negligible natural flow capability except where natural fractures have developed. Many such organic-rich shales can be demonstrated to have been hydrocarbon-source rocks prior to becoming shale-gas reservoirs.

For a shale to be an effective hydrocarbon source rock requires a minimum organic-matter content, measured as TOC, of 2–3 wt%, comprising oil-prone, hydrogen-rich labile kerogen with a Rock-Eval pyrolysis Hydrogen Index (HI) of greater than 250 mg hydrocarbon(HC) $g^{-1}$ TOC. Organic-rich black shales deposited under marine anoxic conditions with a high content of algal and planktonic kerogen form the most prolific oil source rocks, typically with TOCs of 5–10 wt% and HI of 400–600 mg HC $g^{-1}$ TOC. Hydrocarbon generation occurs due to burial-related increases in temperature and pressure, temperatures of 110–120 °C generally being required to generate oil and 130–150 °C for gas generation,[15] occurring at depths of around 4 km and greater than 5–6 km, respectively, in areas with "typical" basinal geothermal gradients of 30–35 °C km$^{-1}$. Expulsion of generated hydrocarbons, oil through light oil, condensate and wet gas to ultimately dry gas, is generally considered to be consequence of the generation process but is not well understood and appears to require a minimum kerogen content of 2.5–3 wt% TOC for a kerogen 'network' to develop.[16] This is believed to allow the hydrocarbons to escape the source rock *via* a potentially hydrocarbon-wet matrix to an adjacent permeable carrier bed where migration to charge a conventional reservoir can occur. This model suggests that the parallel, generally water-wet mineral matrix of the shale source rock may not be involved in the expulsion process, with implications for their potential later development as shale-gas reservoirs. Prior to the widespread development of shale resources, the expulsion of hydrocarbons from the source rock was generally believed to be efficient, values of 60–90% being quoted.[9] However, the large volume of shale gas present in oil-source rocks that have been matured into the gas-window indicates that these values are probably an over-estimate.

The oil and gas reservoired in shales has been demonstrated to represent hydrocarbons not expelled from the source during hydrocarbon generation, with up to a third of generated products being retained.[9] In particular, oil-prone kerogen-rich shales, such as the Barnett, Marcellus, Fayetteville, Haynesville and Eagle Ford Shale formations, are the richest, most prolific shale-gas reservoirs. This can be explained as due to the thermal cracking of bitumens formed as a by-product of oil-generation as well as unexpelled oil remaining trapped in shales by later maturation in the gas-window. This results in the initial generation of light oils and condensates followed by liquids-rich wet gas, with the ultimate generation of large volumes of dry gas continuing through to high levels of maturation (VR of 3–4% Ro). Continued cracking of the residual kerogen components remaining in the shale post oil-generation provide a further source of gas and appear to be critical in providing a considerable proportion of the porosity within the shale. Clearly a proportion of these later-stage maturation products (condensates and dry gas) are expelled to conventional reservoirs, but the prolific shale gas reservoirs show that much remains trapped in the source rock.

## 3.2 Porosity

Determination of porosity in shales requires several techniques to define the parameter at all scales,[17] based on either "traditional" core plugs or on crushed and sieved core as typified by standardised methodology developed by the US Gas Research Institute.[18] On core plugs, mercury porosimetry can be used to characterise pores down to around 5 nm, while smaller pores down to 0.3 nm diameter can be measured by low pressure $N_2$ and $CO_2$ sorption, total porosity determination requiring He porosimetry (see Figure 7).

Despite considerable variation in composition, depositional setting and compaction/maturation history, pore types within shale reservoirs fall into

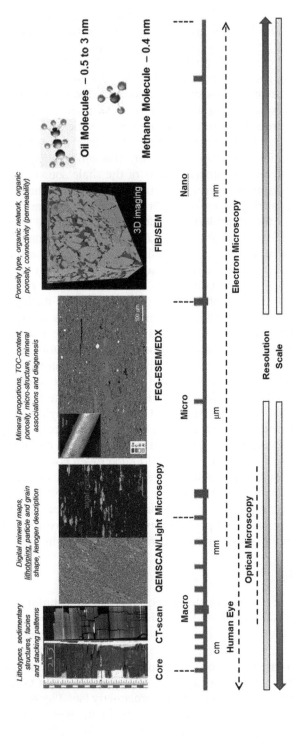

**Figure 7** Porosity scale range and analytical techniques.

three main categories:[19] (1) phyllosilicate framework micropores (defined as >0.75 μm diameter,[20] occurring in both intra- and inter-particle forms[21] and defined by a framework of plate-like clay minerals ranging in size from 5 nm to in excess of 1000 nm); (2) carbonate dissolution micropores (typically located along the margins of calcite and dolomite grains ranging from 50 to more than 1000 nm in size); and (3) organic matter nanopores (defined as <0.75 μm diameter,[20] ranging in size from 10 to several 100 nm within kerogen particles and organo-clay floccules). Micropores also occur in pyrite framboids, within fossils and as micro-cracks. Porosity development within shales can therefore be seen to be complex and highly variable, with the majority of pores of nanometer size, developed predominantly within organic matter in some shales and in the mineral matrix in others, potentially with different reservoir characteristics.[22] This is illustrated by the Barnett Shale, with 20% *inter*-granular porosity and 80% *intra*-kerogen porosity; compared to the Haynesville Shale the pores types are equally distributed. Posidonia shales have 35 to 90% of the total porosity of 5 nm diameter or smaller, located equally between inorganic matrix and organic matter. Clay-rich mudrocks have lower visible porosities than calcareous shales by a factor of 2–3 times, porosity within calcareous nanoplankton and microfossils appearing to account for this.

Most visible porosity (>28 nm) in immature shales occurs within the inorganic matrix,[23] which is lost mostly by compaction during burial through the oil window, although additional loss by in-filling due to bitumen generation appears to play an important role. In Wealden and Posidonia Shales this porosity reduction in the oil window is of the order of about 50%, from about 10–12% to about 4–6%. With increased burial through the gas window, porosity was observed to increase from *ca.* 4–6% to around 10%; crucially, this porosity was observed to be almost entirely associated with organic matter and shows a strong correlation with TOC. Jarvie *et al.*[9] demonstrate that, with no compaction, a 35% carbon loss occurs with the thermal generation of gas from organic matter, calculating a 4.90% porosity increase resulting from maturation of a shale containing 7 wt% TOC.

The discovery of the importance of organic matter porosity is recent,[20,24] enabled by the use of Ar ion milling of shale ultra-thin sections using both focussed ion-beam and broad-beam techniques for high-resolution imaging by scanning electron microscopy (Focussed-Ion-Beam (FIB)-SEM and Broad-Beam (BB)-SEM). In the siliceous Barnett Shale, Loucks *et al.*[20] found that most of the porosity is in the form of nanopores located within grains of organic matter with a median size of 100 nm and size range from 5 nm to greater than 800 nm. Connectivity between pores controls permeability, which is key to gas flow, and in the Barnett, Loucks *et al.*[20] observed long pore throats (200 nm and greater) but with widths less that 20 nm, comparable with observations derived from capillary-pressure studies on other shales.[25,26] A strong relationship with the thermal maturity of the shales was also found, with kerogen grains within low maturity Barnett shales (VR <0.7%) having no internal nanopores, with nanoporosity only evident parallel to kerogen grain boundaries, while high maturity shales (VR >0.8%) have well-developed kerogen nanoporosity as noted above (see Figure 8).

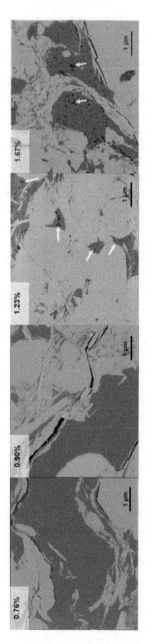

**Figure 8**   SEM shale images showing porosity develop in kerogen only in high maturity shales.

Other workers report similar findings: Modica and Lapierre for the Mowry Shale,[27] while Hover *et al.*[28] found no visible intercrystalline or interparticle matrix porosity in low-maturity Antrim and New Albany shales, while in the Woodford shale Curtis *et al.*[29] found no organic porosity development in shales below VR of 0.9%, with the first appearance of pores occurring with gas window maturity of 1.2% Ro: however, porosity development was variable within samples, indicating that organic matter composition also controls kerogen porosity development. This appears to be confirmed by a study on the Marcellus Shale which found that variation in TOC was a stronger control on organic porosity than thermal maturity,[30] noting that a relationship between TOC and porosity only held at lower TOC values, with bitumen-like material predominating in organically richer shales.

Bitumen, generated as part of the maturation process in the oil window, has been seen to contain nanopores in Marcellus shales with gas window maturity of VR 1.1 and 3.1% Ro,[31] and in the Barnett Shale.[32] Bitumen generation has been suggested as the cause of the reduction in organic porosity in the late oil window, such as recorded in the Posidonia black shales of the Hils Syncline located in the Lower Saxony Basin of Germany where shale porosity, present at lower maturity levels in the oil window, was found to decrease through the late oil window. Maximum porosity development was recorded at gas-window maturities, suggesting that the thermal cracking of these bitumens could be at least partially responsible for this increase in organic porosity.[33] The thermal maturation process also affects the density of the kerogen, which increases with maturity from 1.2 to 1.65 gc m$^{-3}$ for immature Type I and Type II kerogens to greater than 1.65 g cm$^{-3}$ at gas window maturities of VR greater than 2% Ro. This has an effect on shale resource determination as kerogen density, derived from bulk density logs, is a key parameter in porosity determination.

The importance of a kerogen "network" within organic-rich shales with TOC >2.5 wt% is well documented as promoting the expulsion of hydrocarbons into adjacent carrier beds.[16] As much of the porosity, particularly in siliceous shales, is located within the kerogen as a result of thermal maturation, the presence of a kerogen matrix is believed to be critical in the permeability of the shale by providing an efficient route for gas to move from pore to artificial fracture.

## 3.3 Permeability: Methods – Traditional Core Plug and Crushed Core

Although resource shales are low permeability rocks by definition, for them to be effective reservoirs they must have minimal matrix permeability to allow significant hydrocarbon flow after stimulation, although clearly this matrix permeability must not be large enough to allow leakage of the hydrocarbon charge over geological time. This implies low permeability and connectivity between pores in the natural state, allied to high capillarity, particularly in shales where the pore system is water wet. Also, it should be borne in mind that these shales form the cap-rock seals to conventional reservoired hydrocarbon traps. Natural fracture systems, both open and cemented, are also an important feature of overall shale permeability and complicate the measurement and determination of effective connectivity and flow in shale reservoirs. Heterogeneities within the shales also impart major controls; for example, permeability measurements parallel to bedding planes can be up to an order of magnitude greater than those measured perpendicular to bedding.[17] Accurate permeability measurements are required to gain an understanding of the fundamental controls on hydrocarbon flow within shales and, importantly, to derive parameters which can be used in the construction of reservoir models used in the determination of the commerciality of a shale play. These measurements also need to be reproducible and comparable between different laboratories. Permeability measurements have traditionally been measured on core plugs but these can readily be seen to be dependent on many factors, not only on orientation of the plug as discussed above, which often leads to these problems. In an attempt to overcome the difficulties and to produce a standardised, fast method to determine shale permeability, the US Gas Research Institute developed a technique using crushed and sieved core,[18] thus removing the effects of both natural fracturing and rock heterogeneity. However, it should be noted that these may be critical controls on fluid flow within a shale reservoir. Measurements of both porosity and permeability in shales are therefore very dependent on both the analytical methods and the experimental conditions used.

Various flow types can occur in very fine-grained rocks, depending on the nature of the porosity and fracturing:

- Natural fractures and macro-pores tend to have Darcy flow (flow rate proportional to pressure gradient);
- Meso-Pores range from Darcy flow through transitional flow to Knudsen diffusion (molecular slippage along the flow channel walls);
- Micro-pores are restricted to Knudsen diffusion; and
- Hydraulic fractures typified by turbulent non-Darcy flow or Darcy flow.

These observations suggest that while fracture flow, either natural or induced, may dominate early reservoir performance, matrix properties control performance over longer periods.

Non-steady-state permeability measurements show that the state of the core, whether "as received", dried or reservoir/room temperature and pressure, has a major effect on the analytical results, with increased moisture decreasing permeability with increasing effective stress compared to dry samples.[34] The composition of the analytical gas is also critical,[35] with smaller size gas molecules having a greater flow potential. Confining stress and pore pressure have also been shown to be important controls,[36] with permeability decreasing as confining stress increases, which can result from reservoir depletion, while decreasing pore pressure increases the

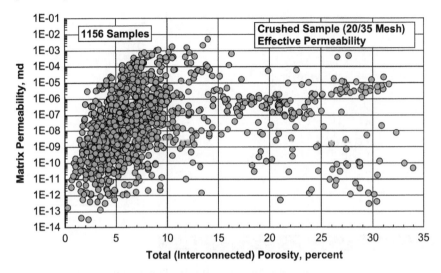

**Figure 9**    Crushed core shale porosity *versus* permeability cross-plot showing the large range of permeability data obtained.
(Modified from Core Laboratories).

effects of Knudsen diffusion and molecular slippage, greatly enhancing permeability.

Analyses of the same samples by both core plug and GRI crushed-core methods show little direct correlation, with the GRI technique producing data with permeability ranges many orders of magnitude lower than the comparable core plugs (see Figure 9). Noticeably, high-end values are missing, indicating that fractures may form an important part in the flow system which is not captured by the crushed shale techniques.

As most pores are associated with organic matter, permeability pathways through the shale connecting these pores to the induced fractures are likely to be greatly influenced, if not controlled, by the three-dimensional arrangement of such organic matter grains within the shale.[20] As noted earlier, development of such an organic matrix within a source rock with TOC $>3.0$ wt% is critical to its hydrocarbon expulsion effectiveness and it appears very likely that a porous organic-matter matrix could greatly enhance the permeability of a shale.[16] Organic matrix permeability would clearly be dependent on the connectivity of the organic pores and, intriguingly, raises the prospect of a potentially hydrocarbon-wet permeability system within gas-shales, separate from the generally water-wet, high capillarity shale mineral matrix.[37]

## 3.4   Gas Storage: Free versus Adsorbed Gas

In shale-gas systems, gas is stored in pores as both adsorbed and free gas[1] (see Figure 10). Adsorbed methane molecules are "attached" to the surfaces

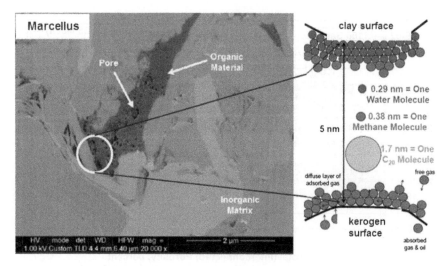

**Figure 10**    Marcellus Shale Formation SEM image showing organic porosity development and pore schematic illustrating free versus adsorbed gas storage.

of the predominantly organic-matter pores and, to a much lesser extent, clay mineral surfaces, while non-adsorbed methane molecules occupy the porosity as free gas. Recent work has shown that the organic matter content of resource shales play a critical role in both hydrocarbon generation and storage, particularly of gas, with experimental data from US lower Palaeozoic-aged shales showing a strong correlation between organic content (TOC) and both microporosity and sorbed methane content.[38]

Sorption experiments have shown that more methane is sorbed by organic matter than by the host shale, with a greater uptake at higher gas-window maturities.[39] Experiments on Jurassic-aged Posidonia shales show that methane sorption occurs predominantly in pores of 5 nm or less diameter, which form between 35 and 90% of the total porosity, and is distributed equally within both the clay mineral matrix and the organic matter. [39] Methane sorption was found to be dependent on organic content (TOC) (see Figure 11) and varies between shales (see Figure 12), decreasing with temperature and increasing with pressure, reaching an equilibrium at pressures of around 10 MPa. Sorption capacity in organic rich shales is dependent on thermal maturity and is reduced by the presence of moisture due to competitive adsorption between water and methane molecules.[39,40]

## 3.5  Mineralogy

Shales with high quartz, carbonate and feldspar contents have a high brittleness and a low Poisson's Ratio and high Young's Modulus, generally exhibiting a good relationship with both natural or artificial fracture development.[41] Successful gas-shales tend to have both a high organic matter content and a "brittle" mineral matrix, indicating that both the generation

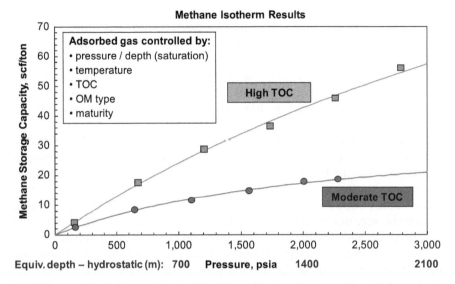

**Figure 11** Methane adsorption relationship with organic matter content.
(Modified from Core Laboratories).

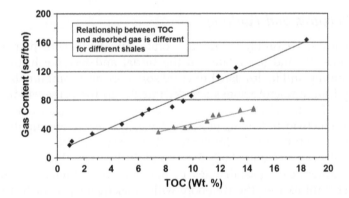

**Figure 12** Variable relationship between methane adsorption and organic matter content for two gas shales.
(Modified from Core Laboratories).

of porosity and gas in the organic matter and the brittle nature of the matrix are critical factors. This is reflected in the ability to hydraulically fracture a shale ("frac'ability") which is related to the fracture strength of the shale and the ability to initiate fractures (high Young's Modulus and low Poisson's Ratio). This is dependent both on the shale matrix composition and the indigenous tectonic stress regime. Brittleness is cited as the key mineralogical parameter in determining the creation of an artificial fracture network connecting the wellbore to the shale micro or nanoporosity,[42] with optimum values of 45% quartz and 27% clay for the Barnett Shale of the Fort Worth

Basin, Texas, USA. In general, optimal shale matrix compositions comprise greater than 50–60% "brittle" minerals, such as silicates and carbonates (typically quartz and feldspar, calcite and dolomite, respectively), and less than 30–40% "ductile" minerals, such as clays and organic matter. Larger contents of ductile clays and organic matter can dramatically reduce the efficiency of the completion process. The ability to initiate fractures and to propagate them through the shale is greatest in quartz or carbonate-rich shales which have overall brittle characteristics, examples include the Mowry, Haynesville and Eagle Ford shales. Shales such as the Marcellus are more clay-rich and therefore more ductile, requiring careful placement of the lateral well in the more clay-poor, less-ductile intervals to maximise fracture initiation and propagation. Very organic-rich and clay-rich "hot shale" units, such as the basal black shale member of the Marcellus Shale, tend to be very ductile and hence difficult to initiate fractures in, although fractures can successfully propagate into and through these units.

Proppants such as quartz sand or artificial ceramic materials are injected into the shales as part of the frac'ing process to prevent closure of the created fractures. Again, mineralogy is important: the use of proppants is most effective in brittle shales, while in ductile shales the proppant grains tend to embed in the fracture walls, allowing eventual closure of the fractures.

## 3.6   Fractures and Faulting

Most shales are naturally fractured,[43] often with joints with perpendicular orientations[44,45] and are related to the burial and tectonic history of the shale. Faulting and fault-related fracture zones are also common. In general, faults and large fracture zones are detrimental to the preservation and exploitation of shale resources as they can allow leakage of gas from a shale or reduce productivity by leaking pressure, typically after tectonic uplift, and can divert and "capture" the hydraulic frac fluids, leaving adjacent rock unfractured. Faults can also potentially form conduits between shale reservoirs and overlying water-bearing sandstones; completion across such a fault zone can allow water influx into the well-bore with consequent loss of hydrocarbon production. Natural fractures are generally beneficial to the completion process,[45,46] with minor open fractures allowing the frac fluids to reach a greater distance into the shale from the well-bore than in an unfractured shale. However, cemented, closed fractures can have to opposite effect as these are weaker than the unfractured shale so fail before the shale matrix, diverting stimulation energy from accessing the shale micro-porosity. This is particularly the case with carbonate-cemented fractures where the bond between the carbonate cement and shale crack wall, generally quartz-rich, is generally weak.[45] Organic-rich "brittle" shales with high quartz or carbonate contents tend to have the greatest development of natural fractures which they observe increases with TOC;[41] as these shales are generally good shale-gas reservoirs, a tentative observation can be made that the presence of natural fractures favours effective completion of shale gas reservoirs.

## 3.7 Confining Elements: "Flow-Unit" Concept

The Flow-Unit is defined as the portion of the shale which has been accessed by the fracking process by the creation of a fracture permeability network, creating a productive reservoir.[47] This is generally a 30–100 m diameter volume of rock located above and along the lateral well due to the effects of buoyancy on the water-based frac fluids. Actual definition of the reservoir volume created by completion (frac'ing) of a well is difficult to determine and is currently under intense investigation. Currently the best determinations are made using micro-seismic measurement of the frac'ing process combined with 3-D seismic data.

Unless the prospective reservoir section is located within a thick section of non-reservoir lean shale, "frac-barriers" are necessary to contain the fractures and to prevent the accidental access of adjacent water-bearing reservoirs. Lean, organic-poor shales can form a frac-barrier as they have different fracture properties from the organic-rich reservoir shales, lacking the organic-matter-based porosity. Typically, such barriers comprise dense limestones or other lithologies with very different fracture initiation properties from the shale, ideally located above and below the prospective reservoir unit. The Onondaga Limestone located at the base of the Marcellus Shale Formation forms such a frac-barrier, preventing the fractures from accessing the underlying water-bearing Oriskany Sandstone. Overlying the Marcellus Formation, the thick organically lean grey shales of the Hamilton/Mahantango formation provide the "top seal".

## 3.8 Resource and Reserve Evaluation

Resources are the amount of hydrocarbons potentially stored within a shale reservoir, while reserves are the proportion of the total which is deemed to be economically producible. Determination of the amount of oil or gas present is, therefore, the critical component of the economic evaluation of a potential shale reservoir. Resource determinations are generally made using a "bottoms up" analytical evaluation of geological parameters.[48] This is based on the mapped area and volume of shale reservoir, the organic matter content (as wt% TOC), mineralogy, water saturation and reservoir pressure. Resources are quoted as either oil in place (Stock Tank Oil in Place or STOIP) or gas in place (GIP), which is calculated using wireline logging measurements of the potential shale reservoir in existing wells penetrating the unit or in logged and cored vertical pilot holes. Data from specifically drilled pilot holes gives the best results as the logging data acquisition programme is tailored to gain optimum results, particularly when calibrated by core data from the reservoir interval. Petrophysical evaluation of the reservoir intervals is based on variations on the Passey method to determine organic matter content *via* the neutron/density logging tool,[49] porosity being derived from this by simultaneous equations utilising kerogen density, with calibration to core-derived data. As noted above, kerogen density increases with maturity, which can introduce significant error in porosity determination if not taken into account. Traditional petrophysical principles are used for water-saturation calculation to determine the hydrocarbon content of the pore space.[50] For gas-shales, the volume of sorbed gas must be taken into account, calculated from the Langmuir volume and pressure at reservoir temperature and pressure conditions and added to the log-derived "free" gas.[51] The Langmuir volume and pressure are derived from the Langmuir isotherm which is derived experimentally for shale reservoir sorption relationships between TOC, maturity and pressure. In general, the Langmuir volume is a function of organic richness and the thermal maturity of the shale, while the Langmuir pressure relates gas adsorption on the organic matter to pressure.

The calculated resource values are typically presented in US "oilfield" units of million barrels of oil (mmbo) per section (square mile) or, in the gas case, billion cubic feet (bcf) gas *per* section. For a shale gas reservoir, resources of greater than 50–60 bcf section$^{-1}$ currently represent a lower economic cut-off for shale gas in the USA, viable economic reservoirs such as the Marcellus Shale having values of 100–180 bcf section$^{-1}$.

In conventional reservoirs, hydrocarbon reserves are derived from resource determinations by the application of a recovery factor derived from modelling of reservoir-flow test data. Clearly, in shale reservoirs with no naturally permeable reservoir this is not possible as the permeability is created post-logging by the stimulation process, which is an unknown until the well is put on production. Potential reserves, known as Expected Ultimate Recovery (EUR), are calculated from the initial production (IP) data and the production decline over a period of time, giving a prediction of the amount of hydrocarbons producible in a given well. This is dependent on the production period and the accuracy increases with time: a problematic scenario, which is solved by the use of standardised decline curves from different shale reservoirs which are related to IP rates and initial production decline rates to give an EUR estimate (see Figure 4). Clearly, choice of analogue decline curve is critical and is potentially the main weakness in resource determinations. Of note is the potential production period of up to 30–40 years predicted by these decline curves, dependent on the mechanical life and economics of the well. Recovery factor is a key parameter to convert resource estimations into technically and economically reserves; this is commonly derived from a combination of shale reservoir parameters, such as mineralogy, and production features such as lateral well length and well spacing. Therefore, not surprisingly, quoted recovery factors show a wide range of 15 to 35% for shale gas,[48] with typical values of 20 to 30%. For example, estimates of recovery factors for US gas and oil shales average 20% for gas with a range of 15–25% and average 4–5% for shale oil with a range of 3–6%.[52]

## 4   Exploration and Exploitation of Shale Reservoirs

Successful exploration for shale resources, whether for shale oil or shale gas, is dependent on the location of good quality organic-rich shale of large regional extent and thickness, and suitable thermal maturity. Present-day

- •••• Gas mature shaly source rock
- ⊕ High richness preferred - 2-3% TOC seems sufficient
- ✳ Fracability – possibility to fracture the shale
- ↕ Minimum thickness – 50-100ft
- ≡ Confining elements
- ↑ Uplift – from 'gas kitchen' to 'economic' drill depth

**Figure 13**  Shale gas exploration geological parameter summary.

depth is also critical, with the shale at an economically drillable depth. While this is variable between basins, in the USA shale gas production is optimal at 3000–4000 m depths; less than 1000–1500 m increases the risk of surface or ground-water-related stimulation issues while depths greater than 4000 m are generally uneconomic, except for prolific highly over-pressured shales such as the Haynesville Shale or the Eagle Ford Shale. Other general parameters are given in Figure 13.

Defining areas of potential shale resources is generally based on pre-liminary screening of existing outcrop geochemical data (TOC, organic richness, maturity, *etc.*), augmented by well log, drill cuttings or core-derived parameters as detailed above, to high-grade areas of potential interest or "sweet-spots". These can be defined as areas where the shale has good or-ganic content, quality and maturity, good "frac'able" characteristics (min-eralogy), thickness and depth. Seismic data are increasingly being used in "sweet-spot" definition, based on seismic attribute evaluation. High organic matter content within a shale affects its seismic character due its low density and low elastic moduli. While there is a linear reduction in bulk density with increasing organic content, seismic velocities show a non-linear reduction. This is complicated by thermal maturity which, as discussed above, in-creases kerogen density, particularly affecting shales within the gas window. Seismic energy propagates through the sub-surface as both P-waves and S-waves which have different properties with variable relationships. Key seismic parameters are P-wave acoustic impedance (AI), which is a function of rock density and P-wave velocity, and the ratio between P-wave and S-wave velocities ($V_p : V_s$ ratio). AI shows a very clear drop with increasing organic content, which can indicate good potential reservoir, while the $V_p : V_s$ ratio (and Poisson's ratio) shows a weak decrease. Data from several resource shales, such as the Marcellus and Eagle Ford, show that the cross-plot of AI *versus* $V_p : V_s$ is a good indicator for shale reservoir potential (see Figure 14),

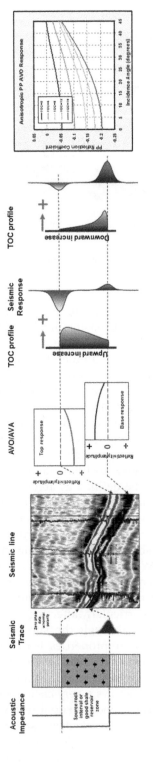

**Figure 14** Shale seismic parameters.

but cannot differentiate between "good" and "not so good" quality reservoir; this is best identified from a cross-plot of bulk density *versus* $V_p$ : $V_s$ ratio. Other combinations of $V_p$, $V_s$ and bulk density, such as Young's Modulus, Lambda-Rho and Mu-Rho, can be used for litho-typing, proxies for geo-mechanical properties ('frac'ability') and potentially also for addressing fluid effects. Seismic velocity within shales is strongly affected by lithological variations and heterogeneities such as laminated organic rich and organic poor units and vertical litho-type variations, leading to high vertical heterogeneity at all scales. In particular, strong vertical anisotropy can affect the interpretation of micro-seismic data.

3-D seismic data are generally used for such shale-quality "sweet-spot" evaluation, but existing good quality 2-D regional seismic data can also be utilised. In general, at shale thicknesses exceeding the seismic tuning effects of the formation boundaries, an organic-rich shale containing gas in particular will have a much lower density than organic-poor or non-gas-bearing shale, which creates a seismic attribute anomaly which can be mapped.[53] Calibration of seismic attributes such P and S-wave anisotropy to well-log and core data allows potentially resource-rich areas of the shale to be determined and mapped, allowing leasing to be optimised and reducing the need for the pilot hole drilling.

Seismic data are also critical in the location of wells, particularly with lateral production wells typically extending 5000–10 000 feet from the vertical pilot well location, as drilling across even very minor faults can lead to lost circulation problems or displacement from the optimal well target horizon, while completion across such faults can result in loss of hydrocarbon production by water influx from a breached cap-rock or seismic activity by fault-plane lubrication; seismic data along a proposed well-bore are therefore critical for geo-hazard determination. Location of a well-pad from which six to eight laterals may be drilled is also critical, with location preferentially in areas of minimal faulting. Vertical pilot wells with detailed wireline logging and core data collection are generally drilled at spacing intervals of around 5 km or when shale parameters have potentially changed, as indicated by logs from non-cored wells or seismic-attribute variation.

Lateral production wells are generally drilled from the pad in opposite directions perpendicular to normal to the direction of maximum horizontal stress, the parallel well-bores generally being 600–800 feet apart. Geological stress is the most important control of fracture growth,[54] with fractures propagating perpendicular to the direction of least principal stress, following the direction of maximum stress. At depths greater than 3000–4000 feet, maximum stress is generally due to overburden, so vertical fracture propagation predominates, the horizontal direction being controlled by tectonic stress direction. Fracture orientation can be obtained from image logging of the reservoir interval of a vertical pilot hole, providing valuable information for determining principal stress directions for drilling the lateral production well.

Successful completion or frac'ing of the lateral wells is very dependent on the characteristics of the shale penetrated. "Brittle" shales (silica or carbonate-rich) have much better completion capabilities and hence flow-rates than ductile, organic-matter/clay-rich shales. Therefore, placement of the lateral in a ductile zone preferably located above an organic-rich layer potentially offers the best production scenario. Hydraulic fracturing of the lateral well-bore is sequentially undertaken, generally at regular intervals separated by movable seals (packers), with up to 20 or more stages *per* well, avoiding any geo-hazards as noted above. However, microseismic monitoring of fracture initiation and propagation during individual completion stages often indicates variable results (see Figure 15), with some intervals showing little fracture activity, suggesting that lateral variations in shale parameters can be significant.[55] Non-completion of such intervals can potentially lead to considerable savings in costs with little effect on overall well productivity.

Currently, completion is largely undertaken using water as the frac medium due the large volumes required for each stage. Minor amounts of chemical additives (typically 0.05% by volume) are added to enhance penetration of the shale structure (surfactants and friction reducers, as well as biocides, scale prevention agents, *etc.*). As noted above, proppants form the other major component of the frac fluid, comprising quartz grains or artificial ceramic material. Flow-back of frac fluids typically occurs during initial production from a well; however, the volumes of fluid produced are typically

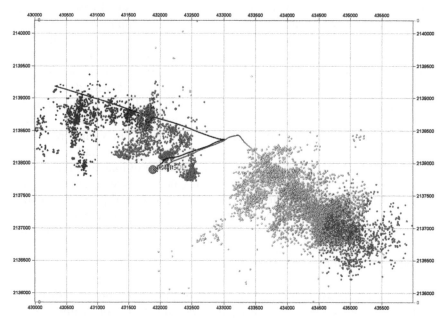

**Figure 15** Micro-seismic data from hydraulic fracture stages on horizontal wellbores.

less than 20–30% of the injected water, particularly in the case of high thermal maturity shales such as the Marcellus.

## 5 Geological/Environmental Considerations

### 5.1 Hydraulic Fracturing (Completion)

Despite the large-scale use of hydraulic fracturing across many areas of the USA and parts of Canada since the early 1990s, there is little hard evidence for the major environmental consequences voiced by many detractors. With the many thousands of wells now completed by hydraulic fracturing, it is likely that incidences of frac-related seismicity and groundwater contamination may have occurred but these appear to be related to individual well or procedural processes rather than to the completion process in general.[56] In many basins around the world, hydrocarbon leakage to the surface and groundwater contamination is an entirely natural occurrence, with gas being derived from high-maturity coals and shallow reservoirs,[57] and is how early explorers located hydrocarbon-bearing basins in the 19th century. Many thousands of wells have been drilled since then, the earliest with only rudimentary casing, if any, and with little or no sealing-off after abandonment. This, along with any uncased faults or fracture systems, may have allowed sub-surface leakage of hydrocarbons from reservoir intervals up into shallower levels, with potential for groundwater contamination. This situation has been known for many years in the old production areas of the USA, such as the Appalachians, but has only come to the forefront with the re-activation of the petroleum industry in this areas with the application of hydraulic fracturing for shale gas extraction,[58] often being cited by detractors of the process as evidence of environmental damage, mostly based on, at best, only circumstantial evidence. While contamination may have occurred, the association with hydraulic fracturing is unproven,[59] and studies are being undertaken to attempt an understanding of this highly contentious issue,[60,61] but the lack of base-line data prior to the start of shale-gas drilling and completion in many basins renders this difficult. Clearly, contamination of groundwater during any stage of shale exploitation, whether by drilling, completion, production or abandonment, is unacceptable and much effort is made to minimise and mitigate this risk.

### 5.2 Groundwater Issues

A simple cross-section of the north-east Appalachian Basin (see Figure 16) illustrates some of the groundwater issues. Most groundwater wells tap shallow gravels and aquifers at depths of a few hundred feet. Beneath this, between depths of around 500 and 2000 feet, lies an interval of generally isolated sands from which early hydrocarbon production took place and now is a potential hazard to drilling. The Marcellus Shale occurs much deeper in the basin at depths of 3500–7000 feet, and the overpressured nature of the

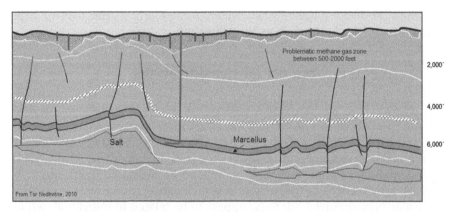

**Figure 16**  Appalachian Basin cross-section showing the relationship between the deep Marcellus Shale Formation reservoir and shallow aquifers.

reservoir indicates sealing from shallower layers by regional seals within the overlying section. This thickness of overburden is more than sufficient to prevent vertical connection between a completed well and near-surface aquifers, as illustrated for the major shale plays in the USA,[54] the largest frac on record attaining a height of 588 m but with a chance of the frac-height exceeding 350 m of around 1%.[56] However, circumstances can be envisaged where this overburden seal may compromised and Figures 17(a) and 17(b) illustrate potential gas leakage scenarios during drilling and completion. In shallow sections by penetration of shallow gas pockets, influx of gas due to lowering pressure by borehole depletion or leakage from deeper sources which could reach shallow aquifers along the borehole or *via* faults penetrated by either the vertical or horizontal sections of the well-bore; these risks can be mitigated by proper drilling procedures with the use of drilling mud of adequate weight and composition and with the subsequent setting of additional surface casing deeper to below any potential aquifer or fault zones, sealed by full cementing. Failure of well casing and abandoned well capping integrity is cited as the most likely cause of well-related gas leakage.[62] Completion of well-bore sections containing faults and natural fracture is a further potential risk and should be avoided; at best, frac'ing such intervals is a waste of frac energy, fluids and proppant, while the risks are much greater as faults can be re-activated as conduits to overlying water-bearing sections which can flow into the well, resulting in loss of the well due to water rather than hydrocarbon production, with the potential risk of contamination of the water-bearing section with drilling and completion fluids and, potentially, gas.

As discussed above, minor faults and fractures zones should be avoided in production well-bores, either vertical or horizontal, as completing across such zones greatly increases the risk of well failure, contamination by leakage of frac fluids or hydrocarbons and the risk of low-level seismic activity. Disposal of fluids into the sub-surface has been implicated in several

**(a)**

Simplified sketch of the basic casing design

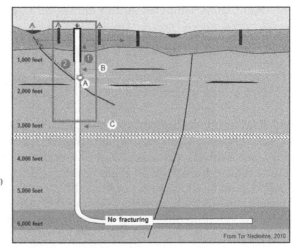

**Methane sources:**
- Penetration of shallow gas pockets (A)
- Seepage due to lowering pressure in formation by bore hole depletion (B)
- Methane sourced from deeper sections (C)

**Migration pathways:**
- Along borehole (1)
- Along faults (2)

**(b)**

Simplified sketch of expanded casing design

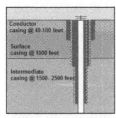

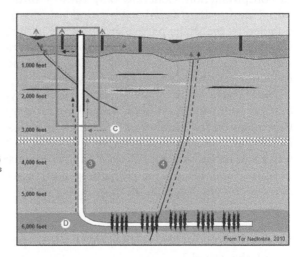

**Methane / injection-water sources:**
- Methane sourced from deeper sections (C)
- Methane and injection-water from Marcellus fracturing (D)

**Migration pathways:**
- Along borehole (3)
- Along crosscutting faults (4)

**Figure 17** Possible risk scenarios for shallow and deep gas leakage from shale gas wells and their mitigation.

cases of seismic activity in the USA[63] and this is believed to be due to lubrication of a fault zone releasing inherent tectonic stresses as minor earthquakes. This may actually be beneficial in areas of high tectonic stress build-up by releasing stress as low level events rather than as a large, potentially damaging earthquake. Lubrication of an unrecognised fault zone during completion of a vertical reservoir section of the Preese Hall #1 well in Lancashire, UK, in 2011 resulted in two minor earthquakes as a result of local tectonic stress release.[64] These risks can be mitigated by the acquisition of a geo-hazard 3-D seismic survey over a real track of the lateral well to identify any faults or fracture zones, with more detailed logging of the lateral well-bore than just the gamma-ray log used for geo-steering to locate any minor sub-seismic resolution faults and fractures.

## 5.3   Completion or "Frac" Fluids

The hydraulic fracturing process, as the name suggests, uses fluids pumped at high pressure in large volumes into the well-bore to initiate and propagate fractures in the shales. These fluids are predominantly composed of water and sand or other proppant (99.95%), with chemicals such as surfactants to reduce surface tension and capillarity effects, corrosion inhibitors and biocides (see Table 1), along with minor amounts of petroleum distillate (light oil) to reduce fluid friction ("slick water"). It should be noted that these additives can be found in many household products and, contrary to popular "belief", includes no "toxic" chemicals.[65,66] After completion of a well, the initial production phase is a clean-up period during which frac-fluid is produced, usually some 70–80% of the injected volume, prior to hydrocarbon production. Generally only 20–30% of the frac-fluid volume is produced during the "flow-back" period prior to production, the rest being retained by the shale, particularly those of high thermal maturity which are very "dry", probably by re-hydration of clay minerals in the shale matrix. The returned frac-fluid is generally cleaned for re-use or treated to remove any remaining chemical traces before disposal.

Water usage by hydraulic fracturing is high, typically 2–6 million US gallons *per* well with 5.6 million US gallons for a typical Marcellus Shale well, dependent on the number of frac stages, and is a critical issue, particularly in areas of restricted supply where it could be in competition with drinking water and irrigation supplies. Recycling and re-use of frac-water is therefore essential, and the use of cleaned-up and recycled water from secondary sources of from desalination may be required.

## 6   Shale Resources – USA *versus* Europe and the Rest of the World

Until the late 2000s, shale resources were predominantly a US phenomenon, with estimates of potential shale gas resources of 2540 trillion cubic feet (TCF),[52] promoted by a buoyant exploration and drilling industry, a well-developed production and pipeline infrastructure and a private leasing and mineral rights ownership virtually unique in the world. The combination of these factors has led to a boom in shale drilling and production, with the result that the USA is likely to become a net gas-exporter later this decade and self-sufficient in oil production. The lack of the combination of these factors elsewhere in the rest of the world probably explains the slow and limited development of shale-based resources.

Worldwide, potential shale gas resources have been identified in many countries (see Figure 18) with estimated reserves of some 7300 TCF and

**Table 1** Typical additives to hydraulic fracturing fluid comprising 0.05% of total volume. Water and quartz sand (proppant) form 99.95% (Chesapeake Energy, South Western Energy, 2014).[65,66]

| Component | Main Compound | Purpose | Common Use |
| --- | --- | --- | --- |
| Acid | Hydrochloric or muriatic acid | Dissolves minerals and initiates rock fractures | Swimming pool cleaner |
| Biocide | Glutaraldehyde | Eliminates bacteria in the water | Disinfectant; medical and dental equipment steriliser |
| Breaker | Sodium chloride | Delays breakdown of gel polymer chains | Table salt |
| Corrosion inhibitor | n,n-dimethyl formamide | Prevents pipework corrosion | Pharmaceuticals, acrylic fibres and plastics |
| Crosslinker | Borate salts | Maintains fluid viscosity as temperature increases | laundry detergents, hand soaps and cosmetics |
| Friction reducer | Petroleum distillate | "Slicks" the water to minimise friction | Cosmetics, hair, make-up, nail and skin products |
| Gelling Agent | Guar gum or hydroxyethyl cellulose | Thickens the water to suspend the proppant | Thickener used in cosmetics, baked goods, ice cream, toothpaste, sauces, and salad dressings |
| Iron control | Citric acid | Prevents precipitation of metal oxides | Food additive; food and beverages; lemon juice *ca.* 7% citric acid |
| Clay Stabiliser | Potassium chloride | Stabilises swelling clay minerals and creates a brine carrier fluid | Used in low-sodium table salt substitute, medicines and IV fluids |
| Oxygen scavenger | Ammonium bisulphite | Oxygen removal from the water to reduce pipework corrosion | Cosmetics, food and beverage processing, water treatment |
| pH adjusting agent | Sodium or potassium carbonate | Maintains effectiveness of components, *e.g.* crosslinkers | Laundry detergents, soap, water softener and dish washers |
| Scale inhibitor | Ethylene glycol | Prevents scale formation | Household cleansers, de-icer, paints and caulk |
| Surfactant | Isopropanol | Increases fracture fluid viscosity | Glass cleaner, multi-surface cleansers, antiperspirant, deodorant, hair colour |

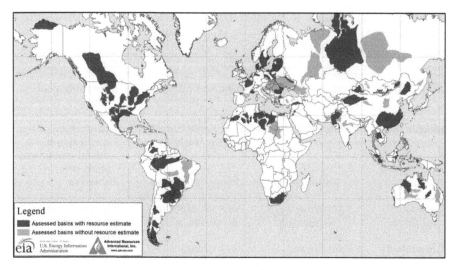

**Figure 18**  Map showing potential for world-wide development of shale gas. (Source: United States basins from U.S. Energy Information Administration and United States Geological Survey; other basins from Advanced Resources International, Inc., based on data from various published studies).[52]

shale oil resources of 345 billion bbls.[52] In the UK, an initial overview of potential shale gas resources gave an estimate of 5.2 TCF;[67] subsequent evaluation of the Carboniferous Bowland Shale in the north of England gave a much larger resource range of 164–447 TCF.[68] Elsewhere, shale exploration and production is taking place in countries such as Poland, Canada, China, Argentina and Australia with variable success, but in others environmental concerns have resulted in drilling bans and moratoria on shale exploration in South Africa (Karoo Basin), France, Germany and Bulgaria to name a few, with strong opposition in other countries such as the UK and Romania. Coupled with these over-riding environmental concerns, other than Canada and potentially Argentina, most countries do not have a thriving, low-cost drilling industry or available infrastructure which greatly increases the cost of the intensive drilling and production facilities required for shale resource development. However, the high energy-price environment present in many of these countries, unlike the USA, can offset the higher drilling and production costs, if the environmental concerns can be overcome, allowing potential reserves in many areas to be exploited.

## 7   Conclusions

The almost overlooked development of shales as gas reservoirs in the 1980–1990s exploded into the phenomenon of shale gas, firstly in the USA and, later, expanding to the rest of world.

Shale resources comprising both gas and oil reside in rocks which in conventional petroleum systems form source rocks and cap rocks/seals and show a strong relationship with the organic-matter content of these shales. Exploration and exploitation techniques are very different from those developed over the last 150 years for conventional resources and require a new "mind-set". In particular, the production of these resources is entirely dependent on the creation of an artificial reservoir by hydraulic fracturing, with the inherent over-hyped and often erroneous environment concerns voiced over the process. While concerns over potential climate change due to the extraction of energy from such resources are legitimate and timely, these resources have a valuable place in supplying a less carbon-rich energy source than coal while the energy-hungry human race develops economic replacement, climate-neutral energy sources.

# References

1. S. I. Montgomery, D. M. Jarvie, K. A. Bowker and R. M. Pollastro, Mississippian Barnett Shale, Fort Worth Basin, north-central Texas: Gas-shale play with multi-trillion cubic foot potential, *Am. Assoc. Petrol. Geol. Bull.*, 2005, **89**, 155–175.
2. R. C. Selley, UK shale gas: The story so far, *Mar. Petrol. Geol.*, 2012, **31**, 100–109.
3. J. A. Harper, The Marcellus Shale – and old "new" gas reservoir in Pennsylvania, *Pennsylvania Geol.*, 2008, **38**, 2–13.
4. R. C. Selley, UK shale-gas resources, Petroleum Geology: North-West Europe and Global Perspectives, *Proceedings of the 6th Petroleum Geology Conference*, Petroleum Geology Conferences Ltd, Geological Society, ed. A. G. Dore and B. A. Vining, London, 2005, 707–714.
5. R. E. Zielinski and R. D. McIver, *Resources and Exploration Assessment of the Oil and Gas Potential in the Devonian Shale Gas of the Appalachian Basin*, U.S. Dept. of Energy, Morgantown, 1982.
6. R. J. Hill and D. M. Jarvie, *Am. Assoc. Petrol. Geol. Bull.*, Special Issue: Barnett Shale, 2007, **91**, 4.
7. T. Engelder and G. G. Lash, Marcellus Shale play's vast resource potential creating stir in Appalachia, *Am. Oil Gas Reporter*, May 2008, pp. 77–78, 81–82, 85–87.
8. G. E. King, *Thirty Years of Gas Shale Fracturing: What have we Learned?* Paper SPE 133456, presented at SPE Annual Technical Conference & Exhibition, Florence, Italy, 19–22 September 16, 2010.
9. D. M. Jarvie, R. J. Hill, T. E. Ruble and R. M. Pollastro, Unconventional shale-gas systems: the Mississippian Barnett Shale of north central Texas as one model for thermogenic shale-gas assessment, ed. R. J. Hill and D. M. Jarvie, *Am. Assoc. Petrol. Geol. Bull., Special Issue: Barnett Shale*, 2007, **91**, 475–499.
10. A. M. M Bustin, R. M. Bustin and X. Cui, *Importance of Fabric on Production of Gas Shale*, SPE 114167, Unconventional Reservoirs Conference, Keystone, CO, Feb 1–12, 2008.

11. L. B. Magoon and W. G. Dow, The petroleum system, ed. L. B. Magoon and W. G. Dow, The petroleum system – from source to trap, *Am. Assoc. Petrol. Geol.*, 1994, memoir 60, pp. 3–24.
12. G. E. Claypool, Kerogen conversion in fractured shale petroleum systems, *Am. Assoc. Petrol. Geol. Bull.*, 1998, **82**(13 supplement), 5.
13. A. M. Martini, L. M. Walter, T. C. W. Ku, J. M. Budai, J. C. McIntosh and M. Schoell, Microbial production and modification of of gases in sedimentary basins: A geochemical case study from a Devonian shale gas play, Michigan Basin, *Am. Assoc. Petrol. Geol. Bull.*, 2003, **87**, 1355–1375.
14. G. Mortis, Horizontal wells show promise in Barnett Shale, *Oil Gas J.*, *January*, 2004, **19**, 58.
15. B. P. Tissot and D. H. Welte, *Petroleum Formation and Occurrence*, Springer, New York, 2nd edn, 1984.
16. M. D. Lewan, Experiments on the role of water in petroleum formation, *Geochim. Cosmochim. Acta*, 1997, **61**, 3691–3723.
17. Y. Yang and A. C. Aplin, A permeability-porosity relationship for mudstones, *Mar. Petrol. Geol.*, 2010, **27**, 1692–1697.
18. Gas Research Institute, *Development of Laboratory and Petrophysical Techniques for Evaluating Shale Reservoir Final Report* (GRI-95/0496), 1996.
19. J. Schieber, *Common Themes in the Formation and Preservation of Intrinsic Porosity in Shales and Mudstones – Illustrated with Examples across the Phanerozoic*, SPE-132370, presented at SPE Unconventional Gas Conference, February 23–25, 2010, Pittsburgh, PA.
20. R. G. Loucks, R. M. Reed, S. C. Ruppel and D. M. Jarvie, Morphology, genesis, and distribution of nanometer-scale pores in siliceous mudstones of the Mississippian Barnett Shale, *J. Sediment. Res.*, 2009, **79**, 848–861.
21. R. G. Loucks, R. M. Reed, S. C. Ruppel and U. Hammes, Spectrum of pore types and networks in mudrocks and a descriptive classification for matrix-related mudrock pores, *Am. Assoc. Petrol. Geol. Bull.*, 2012, **96**, 1071–1098.
22. C. H. Sondergeld, K. E. Newsham, J. T. Comisky, M. C. Rice and C. S. Rai, *Petrophysical Considerations in Evaluating and Producing Shale Gas Tesources*, SPE 131768, presented at SPE Unconventional Gas Conference, Pittsburgh, Pennsylvania, USA, 23–25 February, 2010.
23. M. E. Curtis, R. J. Ambrose, C. H. Sondergeld and C. S. Rai, *Structural Characterization of Gas Shales on the Micro- and Nano-scales*, SPE-137693, presented at CSUG/SPE Canadian Unconventional Resources and International Petroleum Conference, Calgary, Alberta, October 19–21, 2010.
24. F. P. Wang and R. M. Reed, *Pore Networks and Fluid Flow in Gas Shales*, SPE-124253, presented at SPE Annual Technical Conference and Exhibition, New Orleans, LA, October 4–7, 2009.
25. M. E. Curtis,R. J. Ambrose, C. H. Sondergeld and C. S. Rai, *Transmission and Scanning Electron Microscopy Investigation of Pore Connectivity of Gas Shales on the Nanoscale*, SPE144391, presented at SPE Unconventional Gas Conference and Exhibition, TheWoodlands, Texas, June 12–16, 2011.

26. P. H. Nelson, Pore throat sizes in sandstones, tight sandstones and shales, *Am. Assoc. Petrol. Geol. Bull.*, 2009, **93**, 1–13.
27. C. J. Modica and S. G. Lapierre, Estimation of kerogen porosity in source rocks as a function of thermal transformation: Example from the Mowry Shale in the Powder River Basin of Wyoming, *Am. Assoc. Petrol. Geol. Bull.*, 2012, **96**, 87–108; doi: 10.1306/04111110201.
28. V. C. Hover, D. R. Peacor and L. M. Walter, 1996, Relationship between organic matter and authigenic illite/smectite in Devonian Black Shales, Michigan and Illinois Basins, USA, in *Siliciclastic Diagenesis and Fluid Flow: Concepts and Applications*, ed. L. J. Crossey, R. G. Loucks and M. W. Totten, SEPM, Special Publication, 1996, **55**, 73–83.
29. M. E. Curtis, C. H. Sondergeld, R. J. Ambrose and C. S. Rai, Micro-structural investigation of gas shales in two and three dimensions using nanometer-scale resolution imaging, *Am. Assoc. Petrol. Geol. Bull.*, 2012, **96**, 665–677.
30. K. L. Milliken, M. Rudnicki, D. N. Awwiller and T. Zhang, Organic matter-hosted pore system, Marcellus Formation (Devonian), Pennsylvania, *Am. Assoc. Petrol. Geol. Bull.*, 2013, **97**, 177–200.
31. M. E. Curtis, R. J. Ambrose, C. H. Sondergeld and C. S. Rai, *Investigation of the Relationship between KerogenPporosity and Thermal Maturity in the Marcellus Shale*, SPE-144370, presented at the SPE Unconventional Gas Conference and Exhibition, June 12–16, 2011, The Woodlands, Texas.
32. S. Bernard, R. Wirth, A. Schreiber, H. -M. Schulz and B. Horsfield, Formation of nanoporous pyrobitumen residues during maturation of the Barnett Shale (Fort Worth Basin), *Int. J Coal Geol.*, 2012, **103**, 3–11.
33. S. Bernard, B. Horsfield, H. -M. Schulz, R. Wirth, A. Schreiber and N. Sherwood, Geochemical evolution of organic-rich shales with increasing maturity: a STXM and TEM study of the Posidonia Shale (Lower Toarcian, northern Germany), *Mar. Petrol. Geol.*, 2012, **10**, 70–89.
34. A. Ghanizadeh, A. Amann-Hildenbrand, M. Gasparik, Y. Gensterblum, B. M. Krooss and R. Littke, Experimental study of fluid transport processes in the matrix system of the European organic-rich shales: II. Posidonia Shale (Lower Toarcian, northern Germany), *Int. J. Coal Geol.*, 2014, **123**, 20–33.
35. F. Javadpour, D. Fisher and D. Unsworth, Nanoscale gas flow in shale gas sediments, *J. Canad. Petrol. Technol.*, 2007, **46**(10), 55e61.
36. R. Heller, J. M. Vermylen and M. Zoback, Experimental investigation of matrix permeability of gas shales, *Am. Assoc. Petrol. Geol. Bull.*, in press.
37. Q. R. Passey, K. M. Bohacs, W. L. Esch, R. Klimentidis and S. Sinha, From oil-prone source rock to gas-producing shale reservoir – geologic and petrophysical characterization of unconventional shale-gas reservoirs, SPE-131350, presented at the CPS/SPE International Oil & Gas Conference and Exhibition in Beijing, China, June 8–10, 2010.
38. D. J. K. Ross and R. M. Bustin, The importance of shale composition and pore structure upon gas storage potential of shale gas reservoirs, *Mar. Petrol. Geol.*, 2009, **26**, 916–927.

39. M. Gasparik, R. P. Bertie, Y. Gensterblum, A. Ghanizadeh, B. M. Krooss and R. Littke, Geological Controls on the Methane Storage Capacity in Organic-Rich Shales, *Int. J. Coal Geol.*, 2013, **123**, 34–51.

40. T. Zhang, G. S. Ellis, S. C. Ruppel, K. Milliken and R. Yang, Effect of organic-matter type and thermal maturity on methane adsorption in shale-gas systems, *Organic Geochem.*, 2012, **47**, 120–131.

41. W. Ding, C. Li, C. Li, C. Xu, K. Jiu, W. Zeng and L. Wu, Fracture development in shale and its relationship to gas accumulation, *Geosci. Front.*, 2012, **3**, 97–105.

42. K. A. Bowker, Recent development of the Barnett Shale play, Fort Worth Basin, *West Texas Geol. Soc. Bull.*, 2003, **42**, 1–11.

43. J. B. Curtis, Fractured shale-gas systems, *Am. Assoc. Pet. Geol. Bull.*, 2002, **86**, 1921–1938.

44. T. Engelder, G. G. Lash and R. Uzcategui, Joint sets that enhance production from Middle and Upper Devonian gas shales of the Appalachian Basin, *Am. Assoc. Petrol. Geol. Bull.*, 2009, **93**, 857–889.

45. J. F. W. Gale, R Reed and J. Holder, Natural fractures in the Barnett Shale and their importance for hydraulic fracture treatments, *Am. Assoc. Petrol. Geol. Bull.*, 2007, **91**, 603–622.

46. K. A. Bowker, Barnett Shale gas production, Fort Worth Basin: issues and discussion, *Am. Assoc. Petrol. Geol. Bull.*, 2007, **91**, 523–533.

47. R. Aguilera, *Flow Units: From Conventional to Tight Gas to Shale Gas to Tight Oil to Shale Oil Reservoirs*, Paper SPE 165360, presented at SPE Meeting, Monterey, California, USA, 19–25 April, 2013.

48. C. McGlade, J. Speirs and S. Sorrell, Methods of estimating shale gas resources – Comparison, evaluation and implications, *Energy*, 2013, **59**, 116–125.

49. Q. R. Passey, S. Creaney, J. B. Kulla, F. J. Moretti and J. D. Stroud, A practical method for organic richness from porosity and resistivity logs, *Am. Assoc. Petrol. Geol. Bull.*, 1990, **74**(12), 1777–1794.

50. G. E. Archie, Introduction to Petrophysics of Reservoir Rocks, *Am. Assoc. Petrol. Geol. Bull.*, 1950, **34**, 943–961.

51. R. J. Ambrose, R. C. Hartmann and M. Diaz-Campos, Shale Gas-in-Place calculations. Part 1: New pore-scale considerations, *Soc. Petrol. Eng. J.*, 2012, **17**, 219–229.

52. EIA, *Technically Recoverable Shale oil and Shale Gas Resources: An Assessment of 137 Shale Formations in 41 Countries outside the United States*, US Energy Information Administration, Department of Energy, Washington DC, 2013; http://www.eia.gov/analysis/studies/worldshalegas/pdf/overview.pdf (last accessed 06/06/2014).

53. H. Løseth, L. Wensaas, M. Gading, K. Duffaut and M. Springer, Can hydrocarbon source rocks be identified on seismic data? *Geology*, 2011, **39**, 1167–1170.

54. M. K. Fisher and N. R. Warpinski, *Hydraulic-Fracture-Height Growth: Real Data*, Society of Petroleum Engineers, 2012; doi: 10.2118/145949-PA.

55. M. Zoback, S. Kitasei and B. Copithorne, Addressing the environmental risks from shale gas development, Worldwatch Institute, Washington DC, 2012; http://www.worldwatch.org/files/pdf/Hydraulic%20Fracturing %20Paper.pdf (last accessed 06.06.2014).

56. R. J. Davies, S. Mathias, J. Moss, S. Hustoft and L. Newport, Hydraulic fractures: How far can they go? *Mar.Pet. Geol.*, 2012, **37**, 1–6.

57. B. Wilson, Geologic and baseline groundwater evidence for naturally occurring, shallowly sourced, thermogenic gas in northeastern Pennsylvania, *Am. Assoc. Petrol. Geol. Bull.*, 2014, **98**, 373–394.

58. S. G. Osborn, A. Vengosh, N. R. Warner and R. B. Jackson, Methane contamination of drinking water accompanying gas-well drilling and hydraulic fracturing, *Proc. Natl. Acad. Sci. U. S. A.*, 2011, **108**(20), 8172–8176.

59. R. J. Davies, Methane contamination of drinking water caused by hydraulic fracturing remains unproven, *Proc. Natl. Acad. Sci. U. S. A.*, 2011, **108**, E871.

60. A. Vengosh, N. Warner, R. Jackson and T. Darrah, The effects of shale gas exploration and hydraulic fracturing on the quality of water resources in the United States, *Procedia Earth Planet. Sci.*, 2013, 7, 863–866.

61. F. J. Baldassare, M. A. McCaffrey and J. A. Harper, A geochemical context for stray gas investigations in the northern Appalachian Basin: Implications of analyses of natural gases from Neogene-through Devonian-age strata, *Am. Assoc. Petrol. Geol. Bull.*, 2014, **98**, 341–372.

62. R. J. Davies, S. Almond, R. S. Ward, R. B. Jackson, C. Adams, F. Worrall, L. G. Herringshaw, J. G. Gluyas and M. A. Whitehead, Oil and gas wells and their integrity: Implications for shale and unconventional resource exploitation, *Mar. Pet. Geol.*, in press.

63. R. Davies, G. Foulger, A. Bindley and P. Styles, Induced seismicity and hydraulic fracturing for the recovery of hydrocarbons, *Mar. Pet. Geol.*, 2013, **45**, 171–185.

64. Royal Society, *Shale Gas Extraction in the UK: A Review of Hydraulic Fracturing*, Issued: June 2012 DES2597, The Royal Society and The Royal Academy of Engineering, 2012.

65. Chesapeake Energy, *Hydraulic Fracturing Fact Sheet*, 2009; http://www.chk.com/Operations/Process/Hydraulic-Fracturing/Pages/Hydraulic-Fracturing-Fluid.aspx (last accessed 14.03.2014).

66. South Western Energy, *Frac Fluid – What's in it?*, 2014; http://www.swn.com/operations/documents/frac_fluid_fact_sheet.pdf (last accessed 14.03.2014).

67. DECC, *The Unconventional Hydrocarbon Resources of Britain's Onshore Basins – Shale Gas*, DECC Promote website, 2011; https://www.gov.uk/government/uploads/system/uploads/attachment_data/file/66172/uk-onshore-shalegas.pdf (last accessed 16.06.2014).

68. I. J. Andrews, *The Carboniferous Bowland Shale Gas Study: Geology and Resource Estimation*, British Geological Survey for Department of Energy and Climate Change, London, UK, 2013.

# Climate Change Impacts of Shale Gas Production

JOHN BRODERICK* AND RUTH WOOD

## ABSTRACT

The climate change impacts of shale gas are considered from a number of perspectives. When normalised per unit of energy produced, greenhouse gas emissions from production and combustion appear to be comparable to, or marginally higher than conventional sources of natural gas, with direct $CO_2$ from combustion dominating. Substantial uncertainties in such estimates remain, and recent atmospheric studies of methane emissions suggest that on-site measurements and emissions inventories from the US oil and gas industry may be significant underestimates. Shale gas is not a low-carbon energy source, and in the absence of an effective climate regime new gas reserves could have a substantial impact on cumulative $CO_2$ emissions and hence the extent of climate change. The quantity of emissions that will likely cause 2 °C of mean surface temperature rise is very low relative to current emissions and trends, with a restricted time period within which it is prudent to burn natural gas. The position of shale gas as a 'transition fuel' and the relevance of comparison with coal as a fuel source rely upon the assumptions that: (1) it completely displaces the use of an alternative more-carbon-intensive fuel; (2) growth in energy demand does not outpace carbon intensity savings; and (3) it does not jeopardise the development of low- and zero-carbon energy systems. It is not clear that these criteria are currently being met. Without a global

---

*Corresponding author.

---

Issues in Environmental Science and Technology, 39
Fracking
Edited by R.E. Hester and R.M. Harrison
© The Royal Society of Chemistry 2015
Published by the Royal Society of Chemistry, www.rsc.org

carbon cap, the unconstrained use of shale gas is inconsistent with the carbon budgets necessary to avoid dangerous climate change.

# 1  Introduction

Shale gas has been hailed as both a climate change hero and villain. Proponents have argued that it has displaced higher-carbon fossil fuels in electricity generation and transport in the USA, reducing national emissions, whilst there has also been widespread reporting in the popular press that shale gas could have a greater climate change impact than burning coal. Indeed, the International Energy Agency (IEA) concluded in its World Energy Outlook special report on the topic that substantial growth in unconventional gas use in the global energy system would result in very significant levels of climate change, with a rise in temperature of the order of 3.5 °C.[1]

As with any energy technology, there are multiple perspectives to address when considering the climate change impacts of shale gas. These are largely related to the boundaries of the analysis. The following have been considered within the scientific and policy literature, with most attention falling on life-cycle estimates and comparisons of Greenhouse Gas (GHG) emissions:

- Life-cycle climate impacts, normalised per unit of gas, for comparison with other sources of gas, *e.g.* conventional production from the North Sea or Liquified Natural Gas (LNG) imports from the Arabian Gulf;
- Life-cycle climate impacts, normalised per unit of electricity, for comparison with other generation sources, *e.g.* nuclear power or coal;
- Cumulative impact of emissions from shale gas within the global energy system; and
- Indirect investment and policy impacts on the energy system of novel gas reserves.

This chapter provides an overview of research on these issues, identifying the key sources of emissions in shale gas production and situating this within energy systems in order to understand the wider implications for climate change.

# 2  Life-cycle Climate Impact of Shale Gas Production

## 2.1  Sources of Greenhouse Gas Emissions

Estimates of the life-cycle climate impact, colloquially termed the 'carbon footprint', of shale gas have identified multiple sources of emissions from the exploration of a potential basin through to the delivery of natural gas to an end-user (see Table 1). Direct emissions arise from the combustion of fossil fuels for energy, intentional venting of methane for safety reasons, and unintentional releases, termed 'fugitive emissions'. Indirect emissions arise

**Table 1** Overview of emissions sources identified in the scientific literature[8,10,14,21]

| Life cycle stage | Activity | Typical emissions sources |
|---|---|---|
| Extraction | Well construction | Combustion of fossil fuels for drilling machinery on site and transport to site, including exploration and pilot wells. Cement, steel and drilling mud production for wells. Land clearance. |
| | Well stimulation | Combustion of fossil fuels for high pressure pumping of fracking fluids and transport of water and proppant to site. |
| | Well completion | Episodic emissions from flaring or venting of gas during flowback before capture equipment installed at wellhead. |
| | Workovers | Energy and materials for cleaning of wells. Methane releases during workovers and liquid unloading. |
| | Other point source emissions | Wellhead and gathering equipment. |
| | Fugitive emissions | Dispersed leaks from produced-water tanks, pneumatic valves not accounted for as part of other assemblies. |
| | Water treatment | Energy for treatment and disposal of produced water. |
| | Conclusion of production | Energy for disposal of wastes and site remediation. Leakages from abandoned wells. |
| Processing | Acid gas removal | Energy consumed and GHGs released during amine recovery. |
| | Dehydration | Combustion and venting emissions from glycol regeneration. |
| | Fugitive emissions | Dispersed leaks from pumps and pneumatic valves not accounted for as part of other assemblies. |
| | Compressors | Energy consumed and GHG leakages from production compressors. |
| Transport | Pipeline construction | Manufacture and installation of steel piping. |
| | Pipeline compressors | Energy consumed and methane leakages from pipeline compressors. |
| | Pipeline fugitive emissions | Methane leakages through joints. |
| End use | Combustion | $CO_2$ and unburned methane from combustion. |

from the manufacture and transport of drilling equipment, construction material, water, chemical additives and other physical necessities, and the treatment of wastes, including wastewater.

Many of the emissions sources are the same for shale gas as for gas supplied from conventional wells, for instance, compression losses in transmission and distribution. Indeed, the largest component of the life-cycle climate impact, the quantity of $CO_2$ released from combustion of the gas, varies only slightly due to the mixture of alkanes present, which is, in practice, regulated in the UK by National Transmission System gas specifi-cations.[2] Additional GHG emission sources from shale gas lie in the energy required for hydraulic fracturing; the manufacture of chemicals and trans-portation of both water and chemicals to the well site; and their subsequent disposal and direct releases during well completion. Emissions from flow-back, the period following drilling when a mixture of gas, liquid and solids is removed from the well to enable production, are the most substantial of these but may be mitigated to a large extent by 'reduced emissions com-pletions'.[3] Diffuse sources of emissions from sub-surface leakage, con-sidered to be most likely the result of well casing failure, from both conventional and shale wells, have also been identified but not quantified.[4]

## 2.2 Quantitative Estimates of Life-cycle Climate Impacts

There are a number of estimates of the life-cycle climate impact of shale gas available in the literature in addition to those of other energy sources. Before presenting these figures it is worth considering some key issues in their production. Comparison of life-cycle estimates for any product is problem-atic due to differences in system boundaries; metrics chosen to relate dif-ferent atmospheric effects of gases; assumed technology performance; allocation between multiple products; treatment of uncertainties; and data quality.

Choosing appropriate comparators and equivalent functional units is es-sential, but is only a first step. In the case of fossil fuels, this is typically either energy content (*e.g.* joules, J) or electrical energy subsequently gen-erated (*e.g.* kilowatt hours of electricity, kWh(e)). For shale gas, this can make a substantial difference in comparisons with coal due to the much greater efficiency of the conversion technology. A coal-fired electricity plant has a thermal efficiency ranging from 36% (pulverised fuel) to 47% (new supercritical plant), whilst a gas-fired power station ranges from 40 to 60%.[5] Monte Carlo analysis,[6] informed by a number of prior studies,[7–12] suggests that the range in possible efficiency of gas power plants has a greater in-fluence on a 'well-to-wires' life-cycle impact than the total uncertainty in quantities of upstream emissions.

Whilst there have been efforts to harmonise life-cycle impacts for some electricity generation technologies,[13] there is as yet no systematic equivalent for shale gas. As a result, the data are presented in Table 2 as relative statements and ought to be considered with some caution. Furthermore, not

**Table 2**   Estimates of comparative life-cycle climate impact of shale gas.

| Study | Life-cycle climate impact of electricity generated from shale gas relative to: | | Methane GWP |
| | Conventional natural gas | Coal | |
| --- | --- | --- | --- |
| Howarth et al. (2011)[7] | 14 to 19% greater | 18% lower to more than 100% greater | 33 and 105 |
| Jiang et al. (2011)[8] | 3% greater | 20 to 50% lower | 25 |
| Skone et al. (2011)[9] | 3.4% greater | 42 to 53% lower | 25 |
| Stephenson et al. (2011)[10] | 1.8 to 2.4% greater | 30 to 50% lower | 25 |
| Burnham et al. (2011)[11] | 6% lower mean value but statistically indistinguishable | 52% lower | |
| Hultman et al. (2011)[12] | 11% greater | 44% lower | 25 |
| JISEA (2012)[21] | Very similar | Less than half | 25 |
| AEA (2012)[14] | 1 to 8% greater | 41 to 49% lower | 25 |
| MacKay & Stone (2013)[15] | 0.5 to 22% greater | 49 to 53% lower | 25 |

all of the assessments are independent, the work by AEA (2012)[14] and MacKay and Stone (2013)[15] relies upon data presented in a number of the other studies.

GHGs have different radiative properties, lifespans in the atmosphere and interactions with other atmospheric components; a number of metrics are available for comparison of these differing impacts from emissions.[16] Global Warming Potential (GWP) is the most commonly used metric for policy appraisal and life-cycle comparison. It integrates the warming effect of an instantaneous release of gas, relative to carbon dioxide, over a chosen time period, typically 100 years. The use of different time periods and different estimates of GWP can significantly affect the quantitative and qualitative conclusions of comparative emissions-accounting studies.

Carbon dioxide and methane are the main GHGs arising from shale gas production, with methane being the more potent of the two. It has an atmospheric lifetime of 12.4 years and is subsequently oxidised to $CO_2$ though a series of reactions, with indirect effects on tropospheric ozone, stratospheric water and sulfate aerosols.[17] Most studies use the Intergovernmental Panel on Climate Change (IPCC) Fourth Assessment Report, AR4 (2007),[18] estimates of GWP, with a tonne of methane taken to have 25-times the impact of $CO_2$ over a 100-year period. The most recent IPCC report, AR5 (2013),[19] has up-dated the advised values for methane to 28-times on a 100-year basis, if climate-carbon feedbacks are not accounted for, and 34-times if they are. Using these values would tend to increase the impact of upstream emissions relative to combustion emissions and increase the importance of mitigation of these sources.

When ranking shale gas against coal or renewable sources of energy such revision is unlikely to make a substantial difference. However, as can be seen from Table 2, presenting a 20-year comparison period can make a much larger difference and alter rankings between fossil fuel sources which have substantially different profiles. Coal has a significantly greater carbon footprint that is dominated by $CO_2$ from combustion when compared with gas over a 100-year timescale. This ranking is reversed for higher levels of methane leakage if a 20-year comparison is made; it is this choice that accounts for Howarth *et al.*'s (2011) report of higher impacts for gas, not the assumed rates of methane venting.[7,10]

The choice of comparison period is a value judgement, not borne from a scientific principle but related to the period over which one is concerned about impacts.[17] Given that equilibrium climate response depends upon cumulative emissions and is insensitive to the timing or peak rate of emissions, this shorter period does not seem appropriate to energy policy analysis.[20]

Substantial uncertainties remain in the precise life-cycle impact of shale gas production,[6] even in cases where detailed field work has been conducted.[21] The majority of the quantitative shale gas studies identified in Table 2 take a 'bottom-up' approach, *i.e.* identifying and summing point sources from direct measurement or by reference to existing inventories, emissions factors and activity factors. However, a number of recent top-down studies, taking total atmospheric measurements and then attributing these to sources by atmospheric modelling, have found substantial discrepancy with bottom-up inventory-based estimates.[22-24] In one case, a tight oil and gas field with processing facilities, the US Environmental Protection Agency inventory estimate of 1.7% leakage of methane, for the field as a whole, was found to be a substantial underestimate of the 2.3 to 7.7% leakage indicated by atmospheric monitoring.[22] In a second case, leakage of 6.2% to 11.7% was identified with a mean value 1.8 times the record for field.[23] National US emissions inventories have also been found to underestimate total methane emissions by a factor of *ca.* 1.5 to *ca.* 1.7 times, with substantial regional discrepancies suggesting that emissions from fossil fuel extraction and processing could be $4.9 \pm 2.6$-times larger than that recorded in inventories.[24] Further work is required to identify the reasons for this and understand the contribution made by shale gas production. Such discrepancies suggest that greater attention should be paid to monitoring, the process of inventory production and background measurement prior to development.[25] Nevertheless, the broad conclusion that shale gas has a life-cycle climate impact that is within the range of other natural gas sources appears to be robust.

## 3 Shale Gas in the Global Energy System

To assess the impacts of shale gas production on climate change, it is necessary to go beyond a simple assessment of the carbon intensity of the gas, to consider the potential role of shale gas in the global energy system. International agreements on climate change mitigation focus on the avoidance of dangerous climate change,[26] defined and agreed as more than 2 °C increase in global mean surface temperature above pre-industrial levels.[27] This ambition is repeated in EU and UK climate policy.[28] To mitigate climate change to this level, considerable reductions in $CO_2$ and other greenhouse gases from the world's energy system are necessary. If shale gas is to have a role in the world's energy system, it must be consistent with existing mandates for the avoidance of dangerous levels of climate change. This section describes the direct and indirect roles of shale gas in the world's energy system and the associated changes in greenhouse gas emissions that might be expected.

### 3.1   The Significance of the Energy System in Contributing to Climate Change

Greenhouse gases, as defined by the Kyoto Protocol, include carbon dioxide ($CO_2$), methane ($CH_4$), nitrous oxide ($N_2O$), sulfur hexafluoride ($SF_6$) and a group of chemicals called hydrofluorocarbons (HFCs). There are other gases that also trap heat, such as ozone and black carbon, however, it is the gases controlled under the Kyoto Protocol that cause the dominant anthropogenic influence on the atmosphere.[17] Of these Kyoto gases, $CO_2$ emissions dominate, being responsible for 76.7% of global emissions in 2004, the main source being fossil-fuel combustion, which was responsible for 56.6% of global greenhouse gas emissions in the same year.[29] $CO_2$ emissions from fossil-fuel combustion are also the fastest-growing source of GHGs globally, increasing by 50% between 1990 and 2012,[29,30] with an average annual increase during the last decade of 3%.[30] To successfully mitigate climate change, addressing the dominant and growing amount of $CO_2$ from fossil fuel combustion is essential. Thus the impact that shale gas exploitation has on the climate is dependent on the role it plays in the wider energy system and how it contributes towards emissions reductions from fossil fuel combustion.

### 3.2   Cumulative Emissions and Climate Change

The challenge of global mitigation positions shale gas in the context of the limits placed on the amount of greenhouse gas emissions that can be released this century while avoiding a 2 °C increase in global mean surface temperature. When shale gas and other fossil fuels are burnt they release $CO_2$ into the atmosphere, with some of the $CO_2$ being taken up by 'carbon sinks' such as the oceans or vegetation, however, that which remains in the atmosphere (45% of $CO_2$ emissions) remains there for over 100 years. Thus $CO_2$ accumulates in the atmosphere and, as a greenhouse gas, traps heat. The amount of global warming that can be expected has been demonstrated to directly correlate with the accumulated amount of $CO_2$ in the atmosphere.[17,31] Consequently there is a great deal of analysis to estimate the amount of 'cumulative emissions' or an 'emissions budget' that the atmosphere can hold without going beyond a 2 °C increase in global mean surface temperature, or even higher.[31,32] The limit of cumulative emissions that can be released into the atmosphere with a good probability of not causing global warming of 2 °C or more can be found in Table 1. The limit takes into account the ability of carbon sinks to absorb some of the $CO_2$ emitted into the atmosphere and sets aside a proportion of the budget for non-$CO_2$ greenhouse gases, providing the gross total $CO_2$ from fossil fuels and land-use change that can be released into the atmosphere with a good chance of avoiding a 2 °C increase. The limit is based on what can be emitted between the years 2000 and 2050. This timeframe was chosen as it has the most significance to decision-makers today, however, this does not mean that after 2050 fossil-fuel burning can resume. In fact, beyond 2050 the cumulative emissions budget is even lower, and analyses suggest net zero-$CO_2$ emissions are necessary by around 2070 to avoid dangerous climate change.[33]

## 3.3  Fossil Fuels in the Context of Emissions Budgets

In summary, to avoid 'dangerous climate change' the most relevant metric is the cumulative amount of greenhouse gases released during the next century, particularly in the next 35 years when decision-makers have the most immediate influence over emissions. Thus, not only the carbon intensity of a fuel is important but the quantity of it which is used and, therefore, the cumulative amount of $CO_2$ it contributes to the atmosphere.

Table 3 demonstrates that, globally, 40% of the cumulative emissions budget associated with avoiding more than 2 °C of warming has already been used up. Global emissions from fossil-fuel burning, cement manufacturing and land-use change, released between 2000 and 2012, add up to 432 Gt $CO_2$. The International Energy Agency estimate that non-energy-related $CO_2$ sources (*e.g.* from chemical processes, land-use change and deforestation) will release approximately 136 Gt $CO_2$ in the period to 2050.[34] This leaves between 322 and 872 Gt $CO_2$ that can be released into the atmosphere from 2013 to 2050 from the energy sector. The average annual rate of fossil-fuel-related emissions between 2000 and 2011 is 29 Gt $CO_2$ yr$^{-1}$; continuing emissions at this rate would mean the remaining $CO_2$ budget was used up by 2023 (20% probability of exceeding 2 °C temperature increase) or 2042 (50%

**Table 3**  Comparison of the amount of $CO_2$ from fossil fuel and land-use change that can be released into the atmosphere from 2000 to 2050 and emissions to date.

| Global emissions budget to avoid a 2°C global temperature increase | 890 Gt $CO_2$ (20% probability of exceeding 2 °C) | 1000 Gt $CO_2$ (25% probability of exceeding 2 °C)[31] | 1440 Gt $CO_2$ (50% probability of exceeding 2 °C) |
|---|---|---|---|
| Amount emitted 2000–2012.[30] | 432 Gt $CO_2$ | 432 Gt $CO_2$ | 432 Gt $CO_2$ |
| Predicted non-energy $CO_2$ emissions 2012–2050.[34] | 136 Gt $CO_2$ | 136 Gt $CO_2$ | 136 Gt $CO_2$ |
| Remaining $CO_2$ budget for energy 2012–2050. | 322 Gt $CO_2$ | 432 Gt $CO_2$ | 872 Gt $CO_2$ |

probability of exceeding 2 °C temperature increase), with zero budget remaining thereafter. However, rather than stabilising, annual fossil-$CO_2$ emissions are increasing; between 2000 and 2012 the global annual emissions increased by 40%. The challenge, therefore, is significant, both to reverse the upward trend in annual emissions and to reduce year-on-year global emissions by 3–9%, depending on the year at which global emissions peak and the reductions start (see Anderson and Bows (2011) for an examination of alternative pathways and emission reduction rates).[35] These constraints are even more challenging if principles of equity between nations are incorporated. Given that it is developed/OECD nations that are chiefly responsible for the historical emissions of GHGs that have caused climate change to date, and which will persist in the atmosphere for decades to come, a climate regime that does not accord a greater emissions allocation to developing nations is unlikely to be internationally acceptable. When reasonable allowances for future emissions are allocated to developing nations, the budget becomes more constrained for developed nations, with timescales for almost complete decarbonisation of their energy systems necessary well before 2050.[35]

Having quantified the remaining global carbon budget associated with 2 °C, we can now consider the availability of shale gas within this limit. Table 4 shows the existing reserves and resources of fossil fuels in the world and the amount of $CO_2$ that would be emitted if they were burnt without abatement measures. Burning conventional reserves of oil and gas alone would likely use up the entire $CO_2$ budget, before burning any coal at all, or eating into the resources that are at present uneconomic to extract. Shale gas reserves represent between 1.8 and 6.5 times the global budget and, if resources are included as well, there is sufficient shale gas to occupy the budget 5.6 to 17.7-times over.

**Table 4**   Global energy reserves and resources and the associated $CO_2$ emitted if burnt unabated.

|  | Reserves (EJ) | Resources (EJ) | Reserves (Gt $CO_2$) | Resources (Gt $CO_2$) | Total Potential $CO_2$ (Gt) |
|---|---|---|---|---|---|
| Conventional oil | 4900–7610 | 4170–6150 | 349–574 | 297–465 | 665–1009 |
| Unconventional oil | 3750–5600 | 11 280–14 800 | 254–444 | 765–1172 | 1102–1496 |
| Conventional gas | 5000–7100 | 7200–8900 | 271–414 | 391–519 | 684–898 |
| Unconventional gas | 20 100–67 100 | 40 200–121 900 | 1091–3912 | 2182–7107 | 3383–10603 |
| Coal | 17 300–21 000 | 291 000–435 000 | 1510–2125 | 25 395–44 022 | 30 296–44 810 |

Source: Global Energy Assessment Table 7.1.[51] Note that resource data are not cumulative and do NOT include reserves, the GEA also includes additional occurrences which are not included in the figures above; these include >40 000 EJ unconventional oil and >1 000 000 EJ unconventional gas. Conversions from EJ to $CO_2$ use IPCC emission factors for crude oil to estimate conventional oil $CO_2$, shale oil figures to estimate unconventional oil; natural gas emission factors for both conventional gas and unconventional gas and the emission factors for coking coal for coal. The upper and lower emission factors for each fuel have been applied to the upper and lower $CO_2$ estimates, respectively.

It is clear from the figures presented in Tables 1 and 2 that at a global level there is more fossil fuel energy available than can be used in a world that is serious about avoiding dangerous climate change. Numerous analyses have demonstrated that a sizeable proportion of conventional fuel reserves must remain in the ground if globally we are to avoid 2 °C or even 3 °C global warming; *a fortiori*, unconventional fuels should remain unburnt.[36] So why are unconventional fuels being currently exploited and plans being put in place to facilitate their exploitation, given global agreements to avoid dangerous climate change? Putting aside the inertia of current global negotiations on mitigation, in part, while Table 4 presents global availability of fuels, these resources are not endowed uniformly. Regional availability of different fossil fuels and demands differ, creating uneven market dynamics. There are regions of the world where access to conventional gas is more limited, leading to reliance on coal; thus here, shale gas could, with the caveats stated below, play a role. Shale gas has been described, therefore, as a transition fuel, enabling countries and regions currently highly reliant on the more carbon-intensive coal generation to reduce the carbon intensity of their energy system as part of a low-carbon transition. The role of shale gas in a low-carbon transition is, therefore, dependent on where it is used and what it displaces – this is determined at present, to a great extent, by energy markets.

## 4 Shale Gas as a Transition Fuel

### 4.1 Conditions and Evidence to Date

The position of shale gas as a transition fuel relies on the assumptions that: (1) it completely displaces the use of an alternative more-carbon-intensive fuel choice; (2) growth in energy demand does not outpace carbon-intensity savings; and (3) it acts as a bridge between current high-carbon-intensive energy systems based on coal and lower future systems where renewables, carbon capture and storage (CCS) and nuclear energy play a dominant role.

First we consider part (1) and the evidence to date that this condition is met by shale gas exploitation. Much of the potential benefit conveyed by shale gas derives from its potential to reduce the carbon intensity of electricity production. The net effect of displacing coal with shale gas depends very much on the region of the world under consideration and the energy prices therein. In the case of the USA, with no national drivers to reduce coal use, evidence suggests that during the rapid exploitation of shale gas and corresponding price reduction, locally, within the USA, there was some displacement of coal use in electricity generation. In fact, between 2005 and 2012 carbon emissions from energy use in the USA fell by 12%, in part due to a reduction in coal consumption of 25% during the same time frame.[37] However, during the same time frame much of the displaced coal production was sold on global markets, exported and burnt elsewhere. The 32% drop in EU coal import prices between 2005 and 2011 and an increase of US coal exports to the EU of 187% have been attributed to the reduction in the USA's demand for coal.[38,39] While there were local emissions reductions during this time frame, due to the sale and combustion of US coal overseas a net reduction in $CO_2$ cannot be guaranteed, net global reductions are dependent on what the USA's exported coal displaced. Similarly, other countries such as China and India with significant amounts of coal-fired electricity production may in future reduce $CO_2$ through a switch to gas, with the net effect dependent on what happens to the coal that is displaced from their energy system; if shale gas is simply burnt as well as coal, as appears to be the case at present, there is no net global carbon saving.

As alluded to earlier, the impacts of shale-gas use on emissions differs regionally, depending on the incumbent energy mix. In the UK and much of Europe, where coal-fired power stations are being phased out through a combination of air quality legislation and climate policy, such as the Large Combustion Plants Directive and EU Emissions Trading Scheme, shale gas will most likely displace conventional gas, nuclear or renewables and the net impact is to increase $CO_2$ emissions compared to a business-as-usual trajectory. Similarly, a study of the probable end-destination of shale gas extracted in British Columbia, Canada (an area holding half the technically recoverable shale gas resources in Canada,[40] where the incumbent energy system is dominated by hydro-electricity), showed the probable end-destination of shale gas was not to replace coal-fired electricity but for export as LNG or to replace conventional gas for use in neighbouring Alberta's energy-intensive tar sand extraction, perpetuating a positive feedback cycle of fossil fuel emission production, clearly not a 'transition role'. The liquefaction and re-gasification of LNG, together with the transport requirements to the end-user, require additional energy, increasing the life-cycle GHG emissions of the fuel source and diminishing potential savings from replacing coal use and hence the potential role as a transition fuel. In conclusion, there is insufficient evidence to date that condition (1) is being met. Shale gas is not displacing an equivalent amount of coal.

The second condition (2) is that increases in energy demand must not outpace the rate at which carbon intensity of energy supply is reduced. To date, global energy statistics demonstrate the rate of decarbonisation has been outpaced by growth in energy demand. On average a 0.3% annual rate of decarbonisation is observed compared to an increase in global energy use of 2% annually since the industrial revolution and hence a net increase year-on-year of energy-related $CO_2$.[41] A six-fold increase in the decarbonisation rate of global energy demand would be needed to stabilise global energy-related $CO_2$ emissions, a twenty-four fold increase to begin to deliver the minimum rate of decarbonisation necessary to provide a reasonable chance of avoiding more than a 2 °C temperature increase. Could shale gas play a part in accelerating the decarbonisation of the energy system? Much will depend on whether it is burnt as well as or instead of coal, as discussed above, and if it does displace coal, the rate at which it does so. As the emissions intensity of shale gas is only approximately one third to one half that of coal, its contribution to the long-term decarbonisation of global energy demand is limited. Given a global target of approximately 90% reduction in $CO_2$ from fossil-fuel use by 2050 and net zero $CO_2$ emissions by 2070, the time frame within which shale gas can be used unabated is limited.

The third condition (3) for the use of shale gas within a transition pathway is that it leads ultimately to a lower-carbon energy system. Thus, it should not divert investment in the research, development and deployment of energy-demand reduction, renewables, carbon capture and storage, and nuclear technologies. Currently, however, economic analysis of shale gas exploitation in the US suggests its continued development will delay the economic feasibility of nuclear energy in the USA by one or two decades,[42] and similarly, without more stringent climate-change policy in the US, delay deployment of renewables and carbon capture and storage by up to two decades.[43] Such delays are significant when set within the context of global energy scenarios that deliver emissions reductions consistent with avoiding 2 °C warming; these rely on large-scale deployment of these technologies, all of which in practice take decades to build at such scale.

It is also important to consider the time frames of shale gas exploitation in considering its role as a transition fuel. Given the lifespan of gas infrastructure, 30 years for a modern Combined Cycle Gas Turbine (CCGT) power station for example, the window of opportunity for shale gas as a transition fuel is time bound. Building new infrastructure post-2020 would run the risk of it not being used for its full design life and becoming a 'stranded asset' where "environmentally unsustainable assets suffer from unanticipated or premature write-offs, downward revaluations, or are converted to liabilities".[44] In the case of shale gas and, more broadly, fossil-fuel-dependent infrastructure, this is an area of on-going research, particularly with reference to investment-fund holders.[45] Furthermore, if appropriate energy policies are not in place, rather than creating stranded assets inconsistent with a low-carbon future, the continued development of fossil-fuel infrastructure could lock countries into dependency on fossil fuels, creating higher cost barriers for alternative energy sources to compete with.

How this plays out in other regions of the world will depend very much on local markets and governance systems, but it appears that in the only country where shale gas has been exploited to date, the US, it is thus far not acting as a transition fuel. Indeed, the IEA reported in their World Energy Outlook supplement "Are We Entering a Golden Age of Gas?" (2011)[1] that a high-gas-use scenario probably would result in 3.5 °C warming, well beyond what is generally regarded as dangerous climate change. Their Chief Economist, Fatih Birol, therefore commented that "We are not saying that it will be a golden age for humanity – we are saying it will be a golden age for gas".[46]

## 4.2   Future Opportunities?

So what role could shale gas have in a rapidly decarbonising world? For shale gas to be a true transition fuel, any displaced coal must effectively remain in the ground, growth in energy demand must not outpace carbon intensity savings and it must be part of a pathway towards truly low carbon energy sources. How could this be achieved? There are two plausible scenarios where shale-gas exploitation could be consistent with climate change agreements. Firstly, if there were an introduction of a global climate regime with a cap commensurate with 2 °C and/or, secondly, if the associated $CO_2$ were prevented from reaching the atmosphere *via* carbon capture and storage (CSS).

To take the first scenario, an annual or periodic global limit on the amount of $CO_2$ that can be released into the atmosphere is set. Regulated entities, nations, cities or industrial actors, must then be bound by some permitting regime, which may or may not involve carbon trading or taxation. In this system, shale gas could be used within a system where $CO_2$ release is limited to a set global cap. This avoids the indirect emissions implications of the national local measures. However, such a system would require a global agreement, including both developed and developing countries alongside globally agreed administration systems. At present, global negotiations are not sufficiently advanced for this to become a reality in the near- to medium-term future.

The second scenario, perhaps in conjunction with the first, is to prevent the majority of emissions associated with shale gas from reaching the atmosphere. If shale gas is used for combustion, this requires carbon capture and storage technology and a socio-economic system that facilitates its deployment. The principle of CCS technology is ostensibly straightforward: the $CO_2$ is removed from the exhaust stream of a large point source, pre- or post-combustion, and transported to a long-term storage site, such as an old oil or gas field or saline aquifer. The technology has significant potential as many of the individual steps of the process are well established in different industries. The challenge is to assemble all steps together to ensure the long-term safe storage of $CO_2$ and to do it at the gigawatt scale. The technology is perhaps most promising when allied with biomass combustion to deliver net negative emissions.

To date approximately eight large-scale carbon capture plants are in operation, predominantly deployed to recycle the $CO_2$ stripped during the pre-processing of natural gas for use in enhanced oil recovery.[47] No carbon capture and storage (CCS) projects are as yet coupled to electricity generation plant, although two carbon capture plants are under construction (Boundary Dam, Canada, and Kemper County, USA) and intending to capture $CO_2$ for enhanced oil recovery rather than for long-term storage.[47] To put the world on a track to a 2 °C future, the International Energy Agency's 450 Scenario requires CCS technology to capture and provide long-term storage of 2.5 Gt $CO_2$ by 2035, or approximately 870 large-scale CCS plants attached to long-term storage: a build rate of approximately 40 per year between 2014 and 2035.[48] To realise this potential, the build rate of these plants needs to be significantly accelerated, at a pace incompatible with delays for a decade or more. Such a roll-out will require substantial economic incentives and regulatory support.

However, a shale gas plus CCS energy pathway is unlikely to realise very low or zero carbon emissions and so will be restricted in the ultimate potential scale of deployment. Upstream emissions may add a not insignificant penalty of up to 20%, dependent upon the source and transport of the gas.[14,49] This has particular implications for CCS where the capture process itself imposes an energy penalty, requiring more fuel and hence realising greater upstream emission, outside of the capture mechanism. Hammond *et al.* [49] estimate the final emissions intensity of electricity from gas CCS to be approximately 80 g $CO_2$e kWh$^{-1}$, four-to-seven times more than nuclear power.[49,50] Depending upon future grid composition, these quantities of emissions are still likely to be problematic, especially for developed economies given the scale and level of decarbonisation required.

In effect, shale gas exploitation without prerequisite controls is a gamble with high stakes. It may be exploited with the intention of its unabated use and an assumption that the global community will renege on existing commitments. In which case, we all, including the energy industry, suffer the impacts of dangerous climate change. Alternatively, if it is assumed that the global community is serious about avoiding dangerous climate change, it is being exploited with the expectation that either a global carbon cap will be in place in due course or with the concomitant support for mass deployment of CCS technology. Without either of these breakthroughs in policy and technology, investment in shale-gas exploration and associated gas infrastructure is, in effect, an investment in a stranded asset. Without a global carbon cap, the unconstrained use of shale gas is inconsistent with the carbon budgets necessary to avoid dangerous climate change.

# References

1. IEA, *World Energy Outlook 2011 Special Report: Are We Entering A Golden Age of Gas?*, International Energy Agency, Paris, France, 2011.
2. Health and Safety Executive, *Gas Safety (Management) Regulations 1996 (GSMR)*, 1996.
3. F. O'Sullivan and S. Paltsev, *Environ. Res. Lett.*, 2012, 7, 044030.
4. S. G. Osborn, A. Vengosh, N. R. Warner and R. B. Jackson, *Proc. Natl. Acad. Sci. U. S. A.*, 2011, **108**, 8172–8176.
5. POST, *Cleaner Coal, Postnote 253*, Parliamentary Office of Science and Technology, London, UK, 2005.
6. C. L. Weber and C. Clavin, *Environ. Sci. Technol.*, 2012, **46**, 5688–5695.
7. R. W. Howarth, R. Santoro and A. Ingraffea, *Clim. Change*, 2011, **106**, 679–690.
8. M. Jiang, W. M. Griffin, C. Hendrickson, P. Jaramillo, J. VanBriesen and A. Venkatesh, *Environ. Res. Lett.*, 2011, **6**(3), 034014.
9. T. J. Skone, J. Littlefield and J. Marriott, *Life Cycle Greenhouse Gas Inventory of Natural Gas Extraction, Delivery and Electricity Production*, US Dept. of Energy, DOE/NETL-2011/1522, 2011.

10. T. Stephenson, J. E. Valle and X. Riera-Palou, *Environ. Sci. Technol.*, 2011, **45**, 10757–10764.
11. A. Burnham, J. Han, C. E. Clark, M. Wang, J. B. Dunn and I. Palou-Rivera, *Environ. Sci. Technol.*, 2012, **46**, 619–627.
12. N. Hultman, D. Rebois, M. Scholten and C. Ramig, *Environ. Res. Lett.*, 2011, **6**, 044008.
13. *J. Ind. Ecol. (Special Issue: Meta-Analysis of Life Cycle Assessments)*, 2012, **16**(S1).
14. AEA Technology, *Climate Impact of Potential Shale Gas Production in the EU*, 2012.
15. D. J. MacKay and T. J. Stone, *Potential Greenhouse Gas Emissions Associated with Shale Gas Extraction and Use*, Dept. Energy and Climate Change, UK, 2013.
16. G. P. Peters, B. Aamaas, M. T. Lund, C. Solli and J. S. Fuglestvedt, *Environ. Sci. Technol.*, 2011, **45**, 8633–8641.
17. S. Solomon, D. Qin and M. Manning, *IPCC 2013: Summary for Policy-makers*, Cambridge University Press, Cambridge, United Kingdom and New York, NY, USA, 2013.
18. S. Solomon, D. Qin, M. Manning, Z. Chen, M. Marquis, K. Averyt, M. Tignor and H. Miller, *Climate Change 2007: The Scientific Basis: Contribution of Working Group I to the Fourth Assessment Report of the Intergovernmental Panel on Climate Change*, Cambridge University Press, Cambridge, UK, 2007.
19. T. F. Stocker, D. Qin, G.-K. Plattner, M. Tignor, S. K. Allen, J. Boschung, A. Nauels, Y. Xia, V. Bex and P. M. Midgley, *Climate Change 2013: The Physical Science Basis*, Cambridge University Press, Cambridge, UK and New York, USA, 2013.
20. M. R. Allen, D. J. Frame, C. Huntingford, C. D. Jones, J. A. Lowe, M. Meinshausen and N. Meinshausen, *Nature*, 2009, **458**, 1163–1166.
21. J. Logan, G. Heath, E. Paranhos, W. Boyd, K. Carlson and J. Macknick, *Natural Gas and the Transformation of the U.S. Energy Sector: Electricity*, Joint Institute for Strategic Energy Analysis (JISEA), NREL/TP-6A50-55538, NREL: National Renewable Energy Laboratory, Golden, CO, USA, 2012.
22. G. Petron, G. Frost, B. R. Miller, A. I. Hirsch, S. A. Montzka, A. Karion, M. Trainer, C. Sweeney, A. E. Andrews, L. Miller, J. Kofler, A. Bar-Ilan, E. J. Dlugokencky, L. Patrick, C. T. Moore, T. B. Ryerson, C. Siso, W. Kolodzey, P. M. Lang, T. Conway, P. Novelli, K. Masarie, B. Hall, D. Guenther, D. Kitzis, J. Miller, D. Welsh, D. Wolfe, W. Neff and P. Tans, *J. Geophys. Res. Atmos.*, 2012, **117**, D04304.
23. A. Karion, C. Sweeney, G. Petron, G. Frost, R. M. Hardesty, J. Kofler, B. R. Miller, T. Newberger, S. Wolter, R. Banta, A. Brewer, E. Dlugokencky, P. Lang, S. A. Montzka, R. Schnell, P. Tans, M. Trainer, R. Zamora and S. Conley, *Geophys. Res. Lett.*, 2013, **40**, 4393–4397.
24. S. M. Miller, S. C. Wofsy, A. M. Michalak, E. A. Kort, A. E. Andrews, S. C. Biraud, E. J. Dlugokencky, J. Eluszkiewicz, M. L. Fischer,

G. Janssens-Maenhout, B. R. Miller, J. B. Miller, S. A. Montzka, T. Nehrkorn and C. Sweeney, *Proc. Natl. Acad. Sci. U. S. A.*, 2013, **110**, 20018–20022.

25. Environment Agency, *Monitoring and Control of Methane from Unconventional Gas Operations*, DEFRA, UK, 2012.
26. United Nations Framework Convention on Climate Change, *Article 2*, UNFCCC, Geneva, Switzerland & New York, USA, 1992.
27. United Nations Framework Convention on Climate Change, *Conference of the Parties 16*, UNFCCC, Cancun, Mexico, 2010.
28. Council of the European Union, *Presidency Conclusions – Brussels, 22 and 23 March 2005 – IV. Climate Change*, European Commission, Brussels, Belgium, 2005.
29. H.-H. Rogner, D. Zhou, R. Bradley, P. Crabbé, O. Edenhofer, B. Hare, L. Kuijpers and M. Yamaguchi, in *Climate Change 2007: Mitigation. Contribution of Working Group III to the Fourth Assessment Report of the Intergovernmental Panel on Climate Change* ed. B. Metz, O. R. Davidson, P. R. Bosch, R. Dave and L. A. Meyer, Cambridge University Press, Cambridge, UK, 2007.
30. C. Le Quéré, G. P. Peters, R. J. Andres, R. M. Andrew, T. Boden, P. Ciais, P. Friedlingstein, R. A. Houghton, G. Marland, R. Moriarty, S. Sitch, P. Tans, A. Arneth, A. Arvanitis, D. C. E. Bakker, L. Bopp, J. G. Canadell, L. P. Chini, S. C. Doney, A. Harper, I. Harris, J. I. House, A. K. Jain, S. D. Jones, E. Kato, R. F. Keeling, K. Klein Goldewijk, A. Körtzinger, C. Koven, N. Lefèvre, A. Omar, T. Ono, G.-H. Park, B. Pfeil, B. Poulter, M. R. Raupach, P. Regnier, C. Rödenbeck, S. Saito, J. Schwinger, J. Segschneider, B. D. Stocker, B. Tilbrook, S. van Heuven, N. Viovy, R. Wanninkhof, A. Wiltshire, C. Yue and S. Zaehle, *Earth Syst. Sci. Data Discuss.*, 2013, **6**, 689–760.
31. M. Meinshausen, N. Meinshausen, W. Hare, S. C. B. Raper, K. Frieler, R. Knutti, D. J. Frame and M. R. Allen, *Nature*, 2009, **458**, 1158–U1196.
32. B. S. Fisher, N. Nakicenovic, K. Alfsen, J. Corfee Morlot, F. de la Chesnaye, J.-Ch. Hourcade, K. Jiang, M. Kainuma, E. La Rovere, A. Matysek, A. Rana, K. Riahi, R. Richels, S. Rose, D. van Vuuren and R. Warren, in *Climate Change 2007: Mitigation. Contribution of Working Group III to the Fourth Assessment Report of the Intergovernmental Panel on Climate Change*, ed. B. Metz, O. R. Davidson, P. R. Bosch, R. Dave and L. A. Meyer, Cambridge University Press, Cambridge, UK, 2007.
33. G. P. Peters, R. M. Andrew, T. Boden, J. G. Canadell, P. Ciais, C. Le Quere, G. Marland, M. R. Raupach and C. Wilson, *Nat. Clim. Change*, 2013, **3**, 4–6.
34. F. Birol, *World Energy Outlook, 2012*, International Energy Authority Publications, Paris, 2012.
35. K. Anderson and A. Bows, *Philos. Trans. R. Soc. London, Ser. A*, 2011, **369**, 20–44.
36. M. Berners-Lee and D. Clark, *The Burning Question: We Can't Burn Half the World's Oil, Coal and Gas – So How Do We Quit?*, Profile Books, London, 2013.

37. Energy Information Administration, *Short Term Energy Outlook January 2014*, 2014.
38. B. Caldecott and J. McDaniels, *Stranded Generation Assets: Implications for European Capacity Mechanisms, Energy Markets and Climate Policy*, Working Paper, Oxford, UK, 2014.
39. European Renewable Energy Council, *Factsheet Shale Gas and its Impact on Renewable Energy Sources*, Brussels, Belgium, 2013.
40. E. Stephenson and K. Shaw, *Sustainability-Basel*, 2013, **5**, 2210–2232.
41. A. Grubler, T. B. Johansson, L. Mundaca, N. Nakicenovic, S. Pachauri, K. Riahi, H.-H. Rogner and L. Strupeit, in *Global Energy Assessment – Toward a Sustainable Future*, ed. T. B. Johansson, A. Patwardhan, N. A. Nakićenović and L. Gomez-Echeverri, International Institute for Applied Systems Analysis, Cambridge, 2012, pp. 99–150.
42. M. Levi, *Bull. Atom. Sci.*, 2012, **68**, 52–60.
43. H. D. Jacoby, F. M. O'Sullivan and S. Paltsev, *Econom. Energy Environ. Policy*, 2012, **1**, 37–51.
44. A. Ansar, B. Caldecott and J. Tilbury, *Stranded Assets and the Fossil Fuel Divestment Campaign: What does Divestment Mean for the Valuation of Fossil Fuel Assets?*, Smith School of Enterprise and Environment, University of Oxford, Oxford, UK, 2013.
45. B. Caldecott, J. Tilbury and C. Carey, *Stranded Assets and Scenarios – Discussion Paper*, Smith School for Enterprise and the Environment, University of Oxford, Oxford, UK, 2014.
46. R. Harrabin, *Anger over Agency's Shale Report.*, BBC, London, UK, 2012.
47. Global CCS Institute, *The Global Status of CSS 2012*, Canberra, Australia, 2012.
48. IEA, *World Energy Outlook 2013 Special Report: Redrawing the Energy-Climate Map*, International Energy Agency, Paris, France, 2013.
49. G. P. Hammond, H. R. Howard and C. I. Jones, *Energy Policy*, 2013, **52**, 103–116.
50. E. S. Warner and G. A. Heath, *J. Ind. Ecol.*, 2012, **16**, S73–S92.
51. H.-H. Rogner, R. F. Aguilera, R. Bertani, S. C. Bhattacharya, M. B. Dusseault, L. Gagnon, H. Haberl, M. Hoogwijk, A. Johnson, M. L. Rogner, H. Wagner and V. Yakushev, in *Global Energy Assessment – Toward a Sustainable Future*, Cambridge University Press and the International Institute for Applied Systems Analysis, Cambridge, UK, 2012, pp. 423–512.

# The Hydrogeological Aspects of Shale Gas Extraction in the UK

ROBERT S. WARD,* MARIANNE E. STUART AND JOHN P. BLOOMFIELD

## ABSTRACT

The UK may possess considerable reserves of shale gas underlying a significant proportion of the UK, but as yet there has been very little exploratory drilling to confirm the resource potential. The areas likely to be exploited for shale gas are overlain in many areas by aquifers used for drinking water supply and for supporting baseflow to rivers. The vulnerability of groundwater and the wider water environment must therefore be taken very seriously. Experience from the United States suggests that groundwater may potentially be contaminated by extraction of shale gas, both from the constituents of shale gas itself, from the hydraulic fracturing fluids, from flowback/produced water which may have a high content of saline formation water or from drilling operations. A rigorous assessment of the risks is required and appropriate risk-management strategies developed and implemented if the industry is to become established in the UK. It is likely that, due to environmental sensitivities, there will be some locations where shale gas exploitation will be considered unacceptable and this may affect the economic viability of the industry. Because we are still at a very early stage, we can take advantage of experience where things have gone wrong elsewhere and ensure progress is made in a controlled way. We must identify and understand the risks to groundwater from shale gas and establish a fully informed risk management strategy for the

---

*Corresponding author.

---

Issues in Environmental Science and Technology, 39
Fracking
Edited by R.E. Hester and R.M. Harrison
© The Royal Society of Chemistry 2015
Published by the Royal Society of Chemistry, www.rsc.org

industry. We must not look back in 20–30 years and regret not taking the actions we have the opportunity to take now.

## 1  Introduction

With the increasing demand for gas in the UK, declining North Sea gas reserves and the drive for greater energy security, attention is turning to alternative domestic sources of gas. One of these is shale gas. Resource studies for the UK are not yet complete, but early indications are that the UK may possess considerable reserves.[1] However, very little exploration has taken place and so its potential as an economically viable source of gas has yet to be determined. Over the next decade it is expected that exploration activity will significantly increase as a pre-cursor to exploitation and an even greater level of industrial activity.

There are significant technical challenges ahead for an industry that has no track record in the UK and these will apply during both the exploration and exploitation phases. The process of shale gas extraction (which, for the purposes of this chapter, includes both exploration and commercial exploitation) involves accessing gas-rich shale at considerable depth below the ground surface and then hydraulically fracturing ('stimulating' or 'fracking') the rock. This controlled fracturing significantly increases the permeability of the shale by creating fissures and interconnected cracks in material that naturally has extremely low permeability. As a result, gas trapped in the rock is released and can flow into the well and then to the surface.

Hydrogeological considerations play a very important role in shale gas extraction for a number of reasons. Significant volumes of water are required to drill and hydraulically fracture the shale and some of this will need to be sourced from groundwater. The drilling of wells from the surface to depths of typically a kilometre or more requires penetration through geological formations near to the surface that contain freshwater (groundwater). This groundwater may be used as a source of water for drinking, for industry (including food production and agriculture) and for supporting stream flow, groundwater-fed wetlands and their associated ecosystems. The drilling activity itself may impact on the quality of groundwater if not managed effectively, but the greatest concern arises from the potential contamination of groundwater (and the wider environmental impact) by the constituents in the fluid used to hydraulically fracture the shale, the water that returns to the surface after the fracturing operation ('flowback' or 'produced water') and the constituents of the shale gas.

Whilst there are already well-developed regulatory regimes in the UK for hydrocarbon operations and for groundwater protection, it is not yet clear how effective they will be for a new and unproven shale gas industry in the UK. This chapter examines some of the key considerations for shale gas extraction in the UK in relation to groundwater management and protection, taking into account experience elsewhere in the world where the industry is already established.

## 2 Potential Shale Gas Resources and Aquifers in the UK

### 2.1 Potential Shale Gas Source Rocks

The establishment of a successful shale gas industry in the United States has led to many other countries considering the potential within their own territories. The UK is one of these and, with support from the current Government and the introduction of economic incentives, a significant amount of exploration is expected over the coming decade. This exploration is needed ahead of any commercial development, as very little is currently known about how much shale gas can be extracted and whether it can be produced economically.

It was over 20 years ago that it was suggested that the UK may have abundant shale gas reserves, but at that time there was little interest as North Sea gas was still plentiful and the technology to extract the gas was in its infancy. There is now significant interest, with growing recognition that there may be considerable on-shore shale gas potential. Originally it was assumed that this potential was restricted to locations where there had been thermal maturation of organic-rich shales to produce (thermogenic) gas. However, it is now known that (biogenic) shale gas can also be formed by microbiological degradation of organic material irrespective of geological age and burial depth.[2] This finding enhances the UK shale gas resource potential dramatically, making many more rocks potentially prospective.

Potential shale gas source rocks occur in many areas of the UK, including the Carboniferous shales in the Midland Valley of Scotland and across Northern England (Pennine Basin, Stainmore and Northumberland Basin system and Widmerpool Trough), the Jurassic shales in the Wessex and Weald Basins and Lower Palaeozoic shales associates with the Midland Microcraton that extends from Wales in the west to the Thames Estuary in the east.[3] Whilst the UK has an abundance of shale, there is considerable variation in the depth and thickness of each of the shale formations and their full extent is not yet known. The potential shale gas source rocks that are currently being considered in the UK are shown in Table 1.

The first of a series of detailed shale gas resource estimates for UK shales was published in 2013.[1] This focussed on the Carboniferous Bowland-Hodder unit (Bowland Shale) across Northern England. A total 'gas-in-place' resource estimate was made using a 3D-geological model based on over 15 000 miles of seismic profile data integrated with outcrop mapping and information from 64 deep boreholes. The study showed that the Bowland Shale can be divided into an upper and lower unit. The lower unit is structurally more complex than the upper unit, which is considered to be more similar to the Barnett Shale in the United States. The relatively complex lower unit is unlike anything encountered in the United States and it is inferred that there will be unique challenges for the shale gas industry in the UK if it is to be successfully developed.

**Table 1**   UK shales of interest to shale gas extraction (summarised from Andrews[1] and from Harvey and Gray).[4]

| Province | Basin | Source shale | Thickness (m) | Comment |
|---|---|---|---|---|
| Scotland | Midland Valley Basin | Strathclyde Group, Carboniferous | Up to 670 | Immature for oil |
| Central Britain | Bowland Basin Edale Basin Widmerpool Trough Gainsborough Trough Cleveland Basin | Bowland–Hodder Unit, Craven Group, Carboniferous | Typically 150 but reaches 890 | First shale of interest – basins within "gas window" |
| Wessex Weald Province | Wessex Basin Weald Basin | Lias, Lower Jurassic Kimmeridge Clay, Upper Jurassic | Thin beds of oil shale Over 600 in centre | Wytch Farm source rock Immature. Probably biogenic |
| Midlands Microcraton | | Tremadoc Shales, Upper Cambrian | Uncertain | – |
| N E England Province | Northumber- land Trough Stainmore Trough | Yoredale, Carboniferous | Shale units tend to be thin | – |
| South Wales-Bristol Basin | | Marros Group, Carboniferous | Uncertain | Interbedded with thick sandstones |

**Table 2**   Shale gas resource (gas-in-place) estimate for the Bowland Shales.[1]

| | Total shale gas resource estimate (tcf) | | | Total shale gas resource estimate (tcm) | | |
|---|---|---|---|---|---|---|
| | Low (P90) | Central (P50) | High (P10) | Low (P90) | Central (P50) | High (P10) |
| Upper unit | 164 | 264 | 447 | 4.6 | 7.5 | 12.7 |
| Lower unit | 658 | 1065 | 1834 | 18.6 | 30.2 | 51.9 |
| Total | 822 | 1329 | 2281 | 23.3 | 37.6 | 64.6 |

(Units tcf and tcm are trillions of cubic feet and trillions of cubic metres, respectively).

The study applied a statistical approach to assess the resource, which took into account variations in the input parameters. As a result, the gas resource estimates are provided as a range with upper (P10), lower (P90) and median (P50) values (see Table 2).

With the UK shale gas in its infancy and with very little exploration activity to date, it is currently too early to make any reliable estimate of the Technically Recoverable Resource (TRR) or Reserve figure (the proportion of the TRR that is commercially recoverable).

## 2.2 UK Aquifers

The UK obtains around 30% of its public water supply from groundwater, with most of this water abstracted from the principal highly productive bedrock aquifers. Examples include the Chalk of Southern and Eastern England and the Permo-Triassic sandstones in the Midlands. However, it is not only the public supply aquifers that are important. Many tens of thousands of private (drinking water and industrial) supplies abstract from secondary (or moderately productive) aquifers and groundwater plays a vital role in maintaining the baseflow in rivers and supplying water to wetlands and the ecosystems that are dependent on this water. Figure 1 shows the distribution of highly and moderately productive aquifers across the UK. Whilst these may be the most productive aquifers, groundwater in other areas (poorly productive aquifers) may also be locally important for baseflow, wetlands and private water supply.

Two of the biggest concerns related to shale gas extraction are groundwater contamination and over-abstraction of water. As a new industry in the UK, these issues must be taken very seriously. Fortunately, the UK, with highly developed and mature groundwater legislation, management/ protection policies and supporting tools, is in a strong position. This is in contrast to the United States where regulation at the start of their development of shale gas was limited and where groundwater issues are now arising (see later). The UK Government(s) and their environment agencies regulate effectively all potentially polluting industries and, in this context, a nascent shale gas industry will be subject to exactly the same environmental regulation. However, as it is not yet clear how the industry will develop, it is still uncertain what any specific challenges will be in the UK environmental setting. We know from the past that poorly regulated and uncontrolled industrial activity can lead to long-term environmental problems and costly remediation. Examples include the contaminated land associated with former gasworks, fuel stations, mining and waste disposal before effective regulation was brought into force.

Ahead of any new activity that may be potentially polluting, there is a need to fully consider the risks associated with it, both in terms of health and safety to humans and to the environment. Jackson *et al.*[5] identify two areas where research is needed ahead of any development of unconventional gas extraction: baseline monitoring and characterisation of pathways and mechanisms by which contaminants may potentially pollute surface/near surface water resources. In the UK, the British Geological Survey (BGS) has also recognised this as being important and is carrying out baseline monitoring in those areas that have been identified for shale gas exploration[6] and separately is developing (jointly with the Environment Agency) a 3D-model of the spatial relationship between potential shale gas source rocks and the principal aquifers in England and Wales. This work uses the BGS 3D Geological Model of Great Britain[7] and the Aquifer Designation dataset.[8] The full extent of each rock type (shale and aquifer) has been mapped and/or

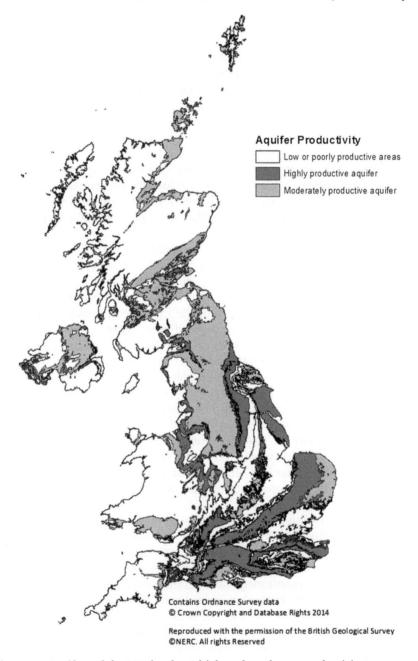

**Aquifer Productivity**

☐ Low or poorly productive areas

■ Highly productive aquifer

▨ Moderately productive aquifer

Contains Ordnance Survey data
© Crown Copyright and Database Rights 2014

**Figure 1**   Aquifers of the UK that have high and moderate productivity.

modelled and the vertical separation calculated for each shale gas source
rock and aquifer pair. The relationship between the different aquifers and
shales is shown in Table 3. Based on this modelling, approximately 30% of

**Table 3**   Relationship of aquifers to potential shale gas targets.[9]

| Aquifer | Shale gas source | Area where both present | Thickness of intervening strata (m) |
| --- | --- | --- | --- |
| Crag | Kellaways/Oxford/Osgodby Clays | East Anglia | 370–460 |
| Crag | Upper Lias | Norfolk | 350–530 |
| Chalk | Kimmeridge/Ampthill Clays | North & South Downs, Hampshire, Wiltshire, Dorset | 50–1200 |
| Chalk | Kellaways/Oxford/Osgodby Clays | Lincolnshire, Norfolk, North & South Downs, Hampshire, Wiltshire, Dorset | 11–1500 |
| Chalk | Upper Lias | Yorkshire, Lincolnshire, Norfolk, North & South Downs, Hampshire, Wiltshire, Dorset | 30–1900 |
| Lower Greensand | Kimmeridge/Ampthill Clays | North & South Downs, Hampshire, Wiltshire, Dorset | 200–1400 |
| Lower Greensand | Kellaways/Oxford/Osgodby Clays | Lincolnshire, Norfolk, Cambridge, Bedford, Bucks, Berks North & South Downs, Hampshire, Wiltshire, Dorset | 37–1800 |
| Lower Cretaceous sandstones-Spilsby | Kellaways/Oxford/Osgodby Clays | Lincolnshire | 250–500 |
| Lower Cretaceous sandstones-Spilsby | Upper Lias | Lincolnshire | 300–600 |
| Corallian | Kellaways/Oxford/Osgodby Clays | Wessex-Weald, Wiltshire | 300–1100 |
| Corallian | Upper Lias | Yorkshire, Wessex-Weald, Wiltshire | 100–1600 |
| Jurassic Oolitic limestones | Upper Lias | Lincolnshire, Norfolk, Cambridge, Bedford, Bucks, Berks North & South Downs, Hampshire, Wiltshire, Dorset | 190–1700 |
| Triassic sandstones | Bowland-Hodder | West Cumbria, Lancashire, Cheshire, Derbyshire, East Midlands, Lincolnshire, Yorkshire | 0–5000 Thinnest in South Midlands and thickest in Cheshire Basin |
| Permo-Triassic sandstones | Upper Cambrian shales | Warwickshire | 0–1300 |
| Permian sandstones | Bowland-Hodder | West Lancashire and Cheshire Basin | 0–5000 |
| Permian (Magnesian) limestone | Bowland-Hodder | Yorkshire, Lincolnshire | 80–1800 |
| None | Marros | NA | NA |
| None | Cambrian Shales | NA | NA |

(a). Extent (full crop) of
potential shale gas source
rocks in England and Wales.

(b). Extent (full crop) of
Principal Aquifers in England
and Wales.

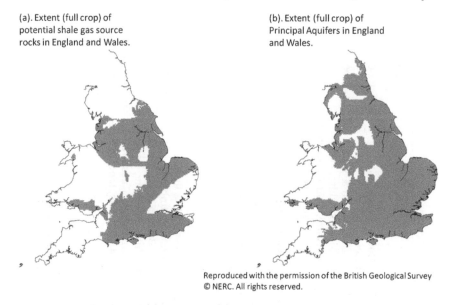

**Figure 2**   Distribution of (a) shale and (b) aquifers across England and Wales.

England and Wales is underlain by potential shale gas source rocks and 50% by geological formations designated as Principal Aquifers (see Figure 2).

The principal groundwater issues thought to be important when considering shale gas exploration and extraction are considered in the following sections. These have arisen from the experience that has emerged from elsewhere in the world, especially the United States where shale gas exploitation is now well established.

## 3   Water Resources

The drilling and completion of shale gas wells can require large quantities of water, as drilling the vertical and horizontal components of the well require water for maintaining hydrostatic pressure, lubrication and cooling of the drill bit, and to return the cuttings to the surface. A further larger volume of water is then needed to carry out the hydraulic fracturing process.

A review of literature associated with shale gas well drilling and stimulation indicates a wide range in the values reported for water use. This variation generally reflects the complexity of the drilling, the geological conditions encountered, total depth of the well and length of the horizontal sections and the number of hydraulic fracturing stages. Estimates of the water requirements for drilling and hydraulic fracturing in different shale gas areas (plays) in the United States, are shown in Table 4.[10] For comparison, the figures quoted by Cuadrilla for the drilling and hydraulic fracturing of their Preese Hall exploratory well in Lancashire are also shown.[11] These figures are lower than those from the United States because they only

**Table 4** Estimated water requirements per well for drilling and fracturing in different shale gas plays (from Mantell[10] and Cuadrilla Resources).[11]

| Shale Play | Drilling ($m^3$) | Fracking ($m^3$) | Total ($m^3$) |
|---|---|---|---|
| Barnett (US) | 950 | 14000 | 14950 |
| Haynesville (US) | 2300 | 19000 | 22300 |
| Fayetteville (US) | 250 | 19000 | 19250 |
| Marcellus (US) | 400 | 21000 | 21400 |
| Eagle Ford (US) | 500 | 23000 | 23500 |
| Bowland Shale (UK)[11] | 900 | 8400 | 9300 |

reflect the water used for a single exploratory borehole rather than multi-stage hydraulically fractured production well.

It is therefore very difficult to estimate how much water will be required in the UK, and in different locations, for shale gas operations because of the significant uncertainty about how the industry may develop and how much it will differ from that elsewhere in the world. The only practical way to consider the potential implications is to examine a number of possible scenarios. This has been done as part of the Strategic Environmental Assessment (SEA) for unconventional gas development in the UK, where a range of between 10 000 and 25 000 $m^3$ has been considered.[12] The SEA considered two activity scenarios: high and low development, each with a number of assumptions being made that resulted in estimates of the total number of production wells ranging from 180 to 2880 and each requiring to be re-fractured once during its lifetime. The resulting range for total water requirement was between 3.6 and 144 million $m^3$.

This water would not all be needed at the same time or in the same location and so an additional major complication is the rate at which the wells would be drilled and hydraulically fractured, and where they will be. Since this information is unavailable as yet, any estimates of water demand for shale gas in the UK are purely speculative. However, if we consider one development scenario proceeding with 100 wells being drilled and completed each year this would require 2.5 million $m^3$, based on the maximum water usage used in the SEA. This is a large number, but how does it compare to how much we already use each year? The most recent UK Government statistics on water abstraction (www.gov.uk) for 2012 estimated that the total non-tidal freshwater (surface water and groundwater) abstraction for England and Wales was 11 700 million $m^3$. This would suggest an overall demand for shale gas equivalent to approximately 0.02% of overall annual abstraction. A breakdown of the most recent water use estimates for England and Wales is shown in Table 5.

Overall shale gas requirements are relatively modest, but the challenge comes from sourcing the water and transporting it to the site at the time required. Whilst some areas of the UK have plenty of water others do not, and there may already be significant stress on water resources in these areas and little, if any, room for additional demand. This is particularly (although

Table 5   Estimated water usage by different users in 2012 (from www.gov.uk).

| Use | Volume (million m³) | Use | Volume (million m³) |
|---|---|---|---|
| Public water supply | 4144 | Spray irrigation | 30 |
| Electricity supply industry | 5702 | Agriculture (excl. irrigation) | 2 |
| Fish farming, cress growing, amenity ponds | 864 | Private water supply | 1 |
| Other industry | 950 | Other uses | 10 |
| **TOTAL (all uses)** | | | **11701** |
| **Estimate maximum annual requirement for shale gas (100 wells *per annum*)** | | | **2.5** |

not exclusively) the case in southern and eastern England. A stark reminder of this was during the drought in 2012 when water use restrictions were introduced.

The environment agencies manage water resources and regularly assess water availability. In England, where demand and pressure on water resources is greatest, a water resource management framework is in place that aims to balance human demand for water with the needs of the environment.[13] Through their catchment abstraction management strategy (CAMS) process the agencies regulate and control both surface water and groundwater abstraction in an integrated way. As with any other industry, the shale gas industry will be subject to these management controls. The most recent assessment of water resources by the Environment Agency has been carried out at a more-local scale than previously. An example of one of the outcomes is shown in Figure 3. This shows for each water body the percentage of time that additional water is available for abstraction. In addition to this, the Environment Agency considers when and how much water may be available for abstraction by considering the relationship between abstraction, river flows and environmental flow needs. To ensure adequate protection of water resources and the environment, resource availability is calculated for different flow conditions between high (Q30) and low (Q95) flows.

The Environment Agency's assessment indicates that there may be significant challenges in sourcing adequate and sustainable quantities of water in some parts of the country where shale gas exploitation is being considered. This will particularly be the case in the South and East of England but, because of local environmental considerations, difficulties cannot be ruled out elsewhere, as Figure 3 shows. The problem is not so great for exploration drilling and testing, but will be significant if industrial-scale development takes place where large numbers of wells will need to be drilled and hydraulically fractured.

**Resource availability - percentage
of time available**

less than 30%
at least 30%
at least 50%
at least 70%
at least 95%

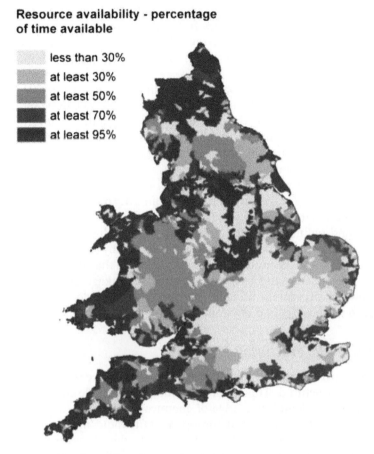

**Figure 3**　Environment Agency water resource assessment for England and Wales.[13]

# 4　Contaminant/Pollutant Sources

Shale gas operations, as with many other industrial activities, involve the use
and/or production of chemicals and materials that are hazardous. The risks
associated with these hazards needs to be assessed and managed effectively
if the activity is to be carried out safely. In managing the risks there is a need
to understand the hazards. The focus here is specifically on risks to
groundwater, but many of them will also be relevant to different parts of the
water environment.

## 4.1　Drilling

Shale gas exploration and exploitation requires the drilling of wells from the
surface to the required depth(s). The process of drilling requires the disturbance of the ground and in many locations the wells will penetrate

freshwater aquifers which supply water for human use and/or baseflow to rivers. Any structure that penetrates freshwater aquifers, such as a well, has the potential to introduce a preferential pathway that could lead to contamination of these water sources if pollutants are allowed to leak or migrate.[14]

Drilling requires the use of water and drilling mud to ensure that the drill bit is lubricated and cooled and so that the drill cuttings can be returned to the surface. During drilling, the drill string and bit will be in contact with the geological formation and groundwater. There is, therefore, a risk of groundwater contamination from the drilling mud and/or mobilisation and transfer of contaminants.

The risks associated with well drilling have to be carefully considered during the planning stage and take into account the purposes of the well, the local surface environment and land use (current and past), the potential receptors and pollutant/environmental exposure pathways and well design. Control measures must be put in place to mitigate any identified risks. These include installation of multiple casings to ensure that different geological horizons are isolated and to act as a barrier to leakage of fluid inside the well; blow-out preventers to avoid over-pressuring in deep boreholes and damage to the casing; and environmental monitoring.

## 4.2   Hydraulic Fracturing Fluids

To optimise the recovery of shale gas hydrocarbon source rock, the shales are hydraulically fractured. This involves pumping large volumes of water containing around 5% and and 0.5% of chemicals in to the well at high pressure. The purpose of the sand (proppant) is to hold open the artificially created fractures and the chemicals to optimise the fracturing process.

The exact composition of the fracturing fluid will depend on the operational conditions, including the geological formation, depth of the well, number of fracturing stages, *etc.* There is no standard recipe and in developing a mix, different chemicals can be used to provide the same function. The number of chemical additives is also not prescribed and so this will vary as well. A representation of the composition of hydraulic fracturing fluid is shown in Figure 4. This identifies some of the key additives that are often used.

The viscosity of fresh water tends to be low, which limits its ability to transport the proppant effectively to achieve a successful fracture stimulation treatment. As a result, some hydraulic fracturing fluids have a gel additive (gellant) to increase the viscosity. Gellant selection is based on the hydrocarbon reservoir formation characteristics, such as thickness, porosity, permeability, temperature and pressure. As temperatures increase, these gels tend to thin dramatically. In order to prevent the loss of viscosity, polymer concentration can be increased (polymer loading) or, instead, cross-linking agents can be added to increase the molecular weight, thus increasing the viscosity of the solution.

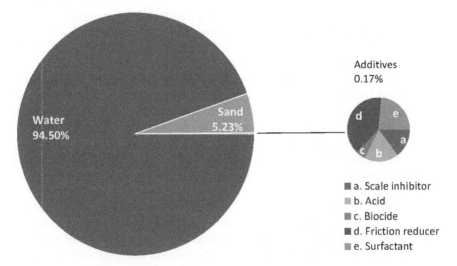

**Figure 4** An illustration of the composition of fracture fluid composition for shale gas well hydraulic fracturing operations.

The fracturing fluid has to reach the bottom of the well, which may be several kilometres below the surface, *via* a relatively narrow diameter (150–200 mm) pipe as efficiently as possible. The addition of friction reducers or surfactants allows the fluids to be pumped to the target zone at a higher rate and lower pressure. Examples of chemicals used as friction reducers are methanol or ethylene glycol.

Acid is utilised in the beginning of the fracture process to clean-up cement that is lodged in the well casing perforations and provide an accessible path to the formation once fracturing fluid is pumped in. Hydrochloric acid is most commonly used at concentrations ranging from 3 to 28%. In stimulations that use acid, a corrosion inhibitor is also often used to hinder the corrosion of steel tubing, well casing, tools and tanks. The addition of 0.1 to 2% of a corrosion inhibitor can decrease corrosion by up to 95%. Concentrations of corrosion inhibitor depend on down-hole temperatures and casing and tubing types. At temperatures exceeding 130 °C, higher concentrations of corrosion inhibitor, a booster or an intensifier may also be necessary. A typical corrosion inhibitor used in shale gas operations is *N,N*-dimethyl formamide.

Biocides are additives that are used to minimise the danger of bacterial corrosion in the wellbore.[15] Fracture fluids typically contain gels that are organic. They can provide an ideal medium for bacterial growth, reducing viscosity and the ability of the fluid to effectively deliver the proppant. Biocides, such as glutaraldehyde are diluted in the fluid in a manner similar to the addition of the corrosion inhibitor.

When a formation contains clay, permeability can be significantly reduced when exposed to water that is less saline than the formation water. As a result, treatment with solutions containing 1 to 3% salt is generally used as a base liquid when clay swelling is probable. Potassium chloride is the most common chemical used as a clay stabiliser due to its ability to stabilise clay against the invasion of water to prevent swelling.

In some operations, a 'breaker' is also added to the fluid in later stages of the process to reduce the viscosity of the gelling agent to help release the proppant from the fluid once the fractures have been created and to increase the volume of flowback water returned to the surface. Chemicals used as breakers may include magnesium peroxide, peroxydisulfate, sodium perborate and glycol. Breakers are normally mixed into the fracturing fluid during the pumping operation to give enough time to transport the proppant to the fracture zone.

At the present time there has only been exploratory drilling (and hydraulic fracturing) for shale gas in the UK, carried out by Cuadrilla at Preese Hall in Lancashire. For this they received approval for a limited suite of chemical additives which comprised: polyacrylamide (friction reducer), salt (fracturing fluid tracer), hydrochloric acid and glutaraldehyde (biocide).[16] At the concentrations used, these were classed as non-hazardous by the Environment Agency. However, this was only for exploratory purposes and operational wells will probably need a different suite of chemicals. These would also need to be approved by the relevant regulator, *e.g.* the Environment Agency.

## 4.3 Flow Back and Produced (Formation) Wastewater

A considerable proportion of the fracturing fluid injected into the well returns to the surface as 'flowback'. Flowback starts immediately and can continue for anything from a few days to a few weeks following hydraulic fracturing. The length of time depends on the geology and geomechanics of the formation. The highest rate of flowback occurs on the first day, and the rate diminishes over time; the initial rate may be as high as 1000 m$^3$ per day.[17] Depending on the geology and extent of fracturing, the volume of produced water may range from between 30 and 70% of the injected fracture-fluid volume.

A certain amount of fluid will continue to emerge from the well over its entire lifetime. This on-going discharge is termed 'produced water' and its composition increasingly reflects that of the geological formation water rather than the injected fluid. The rates of produced-water discharge are generally low and the volumes can be relatively easily handled on the surface. The principal problem is the potentially large volumes of flowback water during the period immediately after hydraulic fracturing.

The composition of flowback water will be similar to the injected fluid(s), modified by the fracturing process and exposure to formation water. It will include the chemicals injected, their transformation and/or breakdown products and formation water. As an example, concentration ranges for the main components of produced water from Marcellus Shales in the United States are shown Table 6. A much larger range of trace elements will also be

**Table 6** Range of constituents in produced water from shale gas wells extracting from the Marcellus Shales, Pennsylvania, US, after Gregory *et al.*[20]

| Component | Concentration range (mg/l) |
|---|---|
| Total dissolved solids | 66 000–261 000 |
| Total suspended solids | 27–3200 |
| Hardness (as $CaCO_3$) | 9100–55 000 |
| Alkalinity (as $CaCO_3$) | 200–1100 |
| Chloride | 32 000–148 000 |
| Sulfate | 0–500 |
| Sodium | 18 000–44 000 |
| Calcium | 3000–31 000 |
| Strontium | 1400–6800 |
| Barium | 2300–4700 |
| Bromide | 720–1600 |
| Oil and grease | 10–260 |

present, many of which can be toxic at elevated concentrations. Examples include arsenic, cadmium and nickel. In addition, shales are hydrocarbon source rocks and so will also contain a range of hydrocarbons and other organics such benzene and naphthalene and other oil petroleum-related compounds. They also may contain relatively high concentrations of uranium and its associated radioactive decay products. This group of materials is referred to as NORM – Naturally Occurring Radioactive Materials – and can lead to a radiological hazard if significant concentrations are contained in the flowback and/or produced waters. Radium-226 and 228 can be present at up to 1000 pCi $g^{-1}$ (pico curies per gramme) potentially many times over the safe disposal limit.[18,19]

The safe handling, storage and disposal/recycling of the wastewater are paramount to avoid risks to humans and the environment. Experience from the United States has highlighted the challenges that could be faced in the UK if large-scale shale gas exploitation was to take place. A number of disposal routes for wastewater are used in the United States, some of which are unlikely to be allowed in the UK due to much stricter regulations. For example, direct discharge to surface water would be prohibited.

In the United States a large proportion of the wastewater is disposed of through deep underground injection. Such an option in the UK is likely to be limited by the availability of suitable locations and environmental regulation. Other disposal routes will, therefore, need to be considered. A number of alternatives have been tried in the US, but none appear to offer a satisfactory solution that could have widespread application in the UK. For example, municipal wastewater treatment plants would not be able to handle the large volumes of highly saline mineralised water as it would damage the biological treatment process. Other forms of treatment, such as reverse/forward osmosis and distillation, are possibilities, but both are energy-intensive processes and the residual waste, although less in volume, would still require disposal. Alternative technologies are actively being researched, but to date none appear to offer an adequate long-term solution. An up-to-date review of treatment technologies is provided by Shaffer *et al.*[21]

An option that offers most promise is to re-use the wastewater for subsequent drilling/hydraulic fracturing operations. The benefits of re-use are self-evident, but until recently re-use has not been seriously considered. This is for a number of reasons, which include the detrimental effect it has on the behaviour and performance of some of the chemical additives in the hydraulic fracturing fluid, leading to poor operational performance of the gas well, and the fact that freshwater is often plentiful and cheap. Increasingly, however, recycling is being considered as regulations are increasingly impacting on disposal options, freshwater resources are becoming limited and technology is advancing.

Evidence of this can be seen in shale gas operations exploiting the Marcellus Shale, Pennsylvania, in the United States. As exploitation has proceeded, wastewater has been disposed of in a range of ways (see Figure 5). Initially there was an increase in treatment at municipal facilities, but this has now returned to pre-exploitation levels. The reduction in use of municipal treatment facilities has been driven by a change in regulation, which means that water can no longer be treated in municipal facilities; similarly, limits on discharges from industrial treatment plants have also made treatment by this method unviable for many plants. In response, there have been sharp increases in the use of deep injection and re-use.[22]

Deep disposal is not permitted in Pennsylvania, so waste water must be transported for considerable distances to use this disposal option and, whilst there has been an increase, there has been a much greater increase in the volume being recycled as this is now considered to be a more practical and cost-effective long-term option.

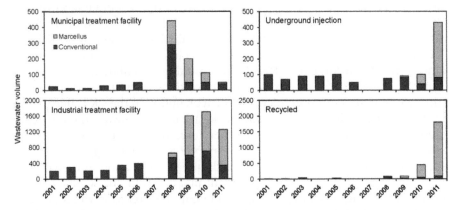

**Figure 5**   Changes in wastewater management methods for conventional wastewater and from Marcellus shale exploitation from Lutz *et al.*[22] (Note differences in vertical scales).

## 4.4 Shale Gas

Shale gas is predominantly, but not exclusively, methane. The source of this gas is the organic material contained in the sediment at the time of deposition. Two processes for the formation of methane are commonly recognised: microbial production (biogenic) and non-biological, chemical production (thermocatalytic or thermogenic).[23] Biogenic gas is formed at relatively shallow depths and low temperatures by the anaerobic microbial decomposition of the organic material. Thermogenic gas is formed at much greater depths and over longer geological timescales through the thermal cracking of the organic material at high temperatures and pressures. Thermogenic gas is usually associated with the major hydrocarbon reservoirs as oil is formed by the same process. The biogenic process does not produce oil.

In general, thermogenic gas has a high methane content with low but significant concentrations of other hydrocarbons such as ethane ($C_2$) and propane ($C_3$), with $C_1/(C_2+C_3) < 100$. In contrast, biogenic gas contains an even higher proportion of methane, with a $C_1/(C_2+C_3)$ ratio between 1000 and 10 000. Where a gas contains such a high proportion of methane it is often described as a 'dry' gas.

Biogenic and thermogenic gases can be readily differentiated and characterised through their geochemistry if the gas has remained close to its source. One method that is often used is the analysis of the stable carbon isotope ratios of the methane ($^{13}C/^{12}C$). $^{12}C$ is the most common isotope but around 1% is $^{13}C$. Due to differences in the biological, chemical and/or physical conditions under which the gas forms, differences in the isotopic ratios occur. The stable isotope ratio ($\delta^{13}C$) values are expressed with reference to an international standard in parts per thousand (permil or ‰). Biogenic methane, on average, contains a greater proportion of isotopically lighter carbon (*i.e.* is more depleted in $^{13}C$) than thermogenic methane. Further differentiation is possible, but one potential problem with relying on stable carbon isotopes is that if the gas migrates from its source it may undergo other changes, such as oxidation, that can result in changes in the ratios and introduce the risk of misinterpretation.[24] The $\delta^{13}C$ of thermogenic methane lies in the range $-110$ to $-55$‰ and biogenic gas in the range $-55$ to $-20$‰.[25]

In addition to methane and other light hydrocarbons, shale gas may also contain small amounts of carbon dioxide, oxygen, nitrogen, hydrogen sulfide, rare or noble gases (argon, helium, neon and xenon) and radon. The concentrations of these gases are generally low and do not present a major hazard, except for radon which is radioactive. This will require appropriate risk assessment and management.

## 5 Contaminant Pathways and Receptors

As an activity that takes place on and below the land surface and involves the use of potential pollutants, there is the potential to pollute groundwater.

This pollution may occur as a result of poor practice, accident or unexpected circumstances that lead to pathways being created between the source of the pollution and the receptor. Receptors in this case include humans, groundwater, surface water and ecosystems. The pathways may be created by:

- Run-off or infiltration to the ground arising from spills and leaks of liquids, chemicals and operational wastewater being transported to/from the site, stored on the site or during use in the hydraulic fracturing process/operation;
- Uncontrolled releases of drilling muds/fluids into non-target geological formations containing groundwater and aquifers during installation of exploration and production wells;
- Migration of drilling muds/fluids or hydraulic fracturing fluids injected at high pressures along natural geological discontinuities (faults and fractures);
- Hydraulic fracturing creating interconnected fractures that extend beyond the intended zone and provide migration pathways for drilling muds/fluids, hydraulic fracturing fluids and formation water to leak into non-target geological formations containing groundwater and aquifers;
- Well failure arising from poor construction or loss of well integrity during the lifetime of operation, including damage resulting from induced seismicity or other ground movement, leading to uncontrolled releases of drilling muds/fluids, hydraulic fracturing fluids and formation waters into non-target geological formations containing groundwater and aquifers; and
- Existing infrastructure such as wells (active or abandoned), mine workings and adits providing pathways for drilling muds/fluids, hydraulic fracturing fluids and formation water to leak into non-target geological formations containing groundwater and aquifers.

Multiple pathways may potentially exist and interact. It is therefore essential that an appropriate level of risk assessment is carried out and effective risk management procedures established. This will require a combination of actions, some of which are engineering-based and will be addressed by ensuring that industry good practice is adopted; others will require site-specific investigation, characterisation and monitoring.

## 5.1   Natural Sub-surface Pathways

Myers[26] identifies two possible mechanisms by which pollutants may migrate from fractured shale to shallow aquifers along natural pathways in the sub-surface: advective transport through the rock matrix, and preferential flow through fractures and other discontinuities. He suggests that there is substantial geological evidence that natural hydraulic gradients can drive contaminants to near the surface from deep evaporite sources. The modelling presented shows it would take tens of thousands of years for pollutants to migrate from depths of >1.5 km to the near surface if transport was through the rock matrix alone. However, hydraulic fracturing of the rock, the introduction of fractures and substantially higher pressure gradients could reduce transport times to tens or hundreds of years under certain conditions. The estimates presented by Myers have been challenged by Saiers and Barth[27] who say the approach used is far too simplistic and neglects critical hydrological factors, such as fluid density gradients, fluid saturation and temperature. They do, however, recognise that there is a need to undertake more modelling to derive estimates of fluid migration and pollutant transport along natural (and induced) pathways from shale gas-bearing rocks.

In the UK there is considerable uncertainty about the geophysical, chemical and hydrogeological properties of the rocks which comprise the natural pathways in the areas that are being considered for shale gas. Knowledge and measurement of sub-surface properties below depths of about 100 m is extremely sparse in all but a very few locations, and so there is an important need to fill this knowledge gap to allow development of sensible conceptual and mathematical models of fluid movement and behaviour in the deep sub-surface before shale gas exploitation proceeds.

## 5.2 Induced Fractures

A frequently expressed concern about shale gas development is that hydraulic fracturing might create fractures that extend well beyond the target formation to overlying shallow groundwater and/or aquifer formations. These would then allow migration of pollutants such as methane, highly saline and mineralised formation water, and fracturing fluids from the target formation to contaminate drinking water supplies.[28]

If hydraulic fracturing is carried out at the depths that are being suggested, *e.g.* greater than 1 km, these fractures would have to propagate considerable distances and, more often than not, through geological sequences that comprise rock types with different physical properties. Because of this, propagation of fractures over these distances is highly unlikely as a result of shale gas hydraulic fracturing operations. A report for New York State concludes that fracking is unlikely to create a pathway beyond the fractured zone and the post-fracking reversal of pressure means that fluids will migrate back to the well.[29]

Work has been carried out to compile information on both natural and shale gas fracture propagation.[30,31] Data were examined from each of the main shale gas formations in the USA and a statistical analysis carried out. The maximum recorded upward propagation of fractures was approximately 588 m and the calculated probability of a fracture extending more than 350 m was around 1%.

## 5.3  Drilling and Well Integrity

The risks associated with well design and drilling have to be carefully considered and take into account the purposes of the well, the geological formations that will be drilled and the materials being used/produced during and after drilling. Any structure that penetrates freshwater aquifers, such as a well, has the potential to introduce a preferential pathway that could lead to contamination if pollutants are allowed to leak.[14] Control measures must be put in place to mitigate any identified risks. These include installation of multiple casing and cement bonded wells, blow-out preventers and environmental monitoring.

The key hazards associated with the drilling operation that could potentially lead to groundwater/surface water contamination are: loss of drilling fluids to the surrounding geological formation(s) (leak-off); well blow-out as a result of gas or fluids under high pressure being encountered in the well bore; and spillages of wastes and chemicals on the surface.

During drilling of the shallow geological formations, drilling fluids which aim to minimise the risk of groundwater contamination are generally used. Examples include the use air and/or water. The well casing and cement that holds it in place provide the seal(s) that is (are) of vital importance both during the drilling phase and then for maintaining the integrity of the well during its lifetime. Failure of the cement or casing surrounding the wellbore poses a significant risk to groundwater. If the annulus is improperly sealed or the seal fails, natural gas, fracturing fluids and formation water(s) containing high concentrations of pollutants may be communicated directly along the outside of the central wellbore between the target formation, drinking water aquifers and layers of rock/groundwater in between.

Studies that have looked at well integrity failure reveal considerable variation in failure rate. Work by Schlumberger estimated that by the time a well is 15-years old there could be a 50% chance of failure.[32] CIWEM reported that 6 to 7% of new wells in Pennsylvania have compromised structural integrity,[33] and a more recent review showed that rates of wells with integrity issues ranged from 2.9 to 75%, with the lower rates reflecting wells drilled since 2010 in Pennsylvania.[46]

Another potential cause of well integrity failure is as a result of ground movement, which includes damage induced by seismic activity triggered by hydraulic fracturing. Casing deformation is relatively common in deep hydrocarbon wells due to geological processes and differences in the properties of adjacent geological formations. This may result in horizontal shearing and subsequent deformation (buckling) of the well casing. At Preese Hall in Lancashire the seismic events which were triggered by the hydraulic fracturing of the exploratory well are also believed to have led to deformation of the well casing as a result of seismically induced rock shear.[34]

## 5.4  Surface Accidental Releases of Liquids and Chemicals

Shale gas well drilling and hydraulic fracturing requires a period of 1–2 months of intense activity around the well, during which spillages or leakage of polluting substances may occur. Activities which have been identified as hazardous include: re-fuelling of diesel tanks, bulk-chemical or fluid transport and storage, equipment cleaning, vehicle maintenance, leaking pipe work, drilling mud/cement mixing areas, wastewater storage and transport. As significant volumes of fluid and chemicals are stored/mixed/used on site there is potential for either direct run-off to drains, ditches and other water courses, or infiltration to ground which may adversely impact surface water quality and ecology or may lead to localised groundwater pollution.

In the United States, storage or retention pits are frequently used for holding freshwater and/or wastewater. These are unlikely to be allowed in the UK and fluids will be required to be held in storage tanks. Tanks can also be used in a closed-loop drilling system. Closed-loop drilling allows for the re-use of drilling fluids and the use of lesser amounts of drilling fluids. Closed-loop drilling systems have also been used with water-based fluids in environmentally sensitive environments in combination with air-rotary drilling techniques. The containment of fluids reduces the risk of leakage and is likely to represent standard practice if shale gas exploitation goes ahead in the UK.

## 6  Risk Assessment, Regulation and Groundwater Protection

There is a well-established oil and gas industry in the UK and more than 2100 wells have been drilled since 1902 for hydrocarbon exploration or exploitation.[35] It is a regulated industry with several regulatory bodies and agencies responsible for the different aspects of the operation (see Table 7). Whilst shale gas is a new development in the UK, many of the regulations and procedures will be applicable or directly transferable. The key bodies that will be involved are the Department of Energy and Climate Change (DECC), the relevant environment agency (Environment Agency, Scottish Environment Protection Agency, Natural Resources Wales or Northern Ireland Environment Agency), Healthy and Safety Executive and local (and mineral) planning authorities. To oversee the safe and responsible development of unconventional oil and gas and ensure co-ordination of activities across the UK, the Government established the Office of Unconventional Gas and Oil (OUGO) in 2013. One of the first outputs from OUGO is a roadmap that provides an introduction to and guidance on the planning and permitting process for unconventional oil and gas exploratory well drilling.[36] OUGO recognises that, as the industry is in its infancy in the UK, the roadmap will need to be revised as legislation develops, new regulations are introduced, or when best practice becomes established.

Currently the roadmap does not address the full range of environmental risks and risk management requirements. For example, it does not cover

*Robert S. Ward, Marianne E. Stuart and John P. Bloomfield*

**Table 7** List of government bodies and agencies with responsibilities for regulating aspects of shale gas exploitation, water supply and environment protection in the UK.

| | Government department/devolved department/agency | | | |
|---|---|---|---|---|
| | England | Wales | Scotland | Northern Ireland |
| Hydrocarbons exploration & development licensing, emissions targets, environmental risk assessment | Department of Energy and Climate Change (DECC) through PEDL (Petroleum Exploration and Development Licenses) rounds | | | Energy Division, Department of Enterprise, Trade & Investment (DETI) |
| Planning permission & access negotiation, environmental impact assessment | Minerals Planning Authority (MPA) Department for Communities and Local Government (DCLG) – Local Planning Authorities | | | Planning & Local Government Group, Department of the Environment (DOE, Northern Ireland) |
| Permission to penetrate coal seams | Coal Authority | | | |
| Notification of drilling operations, well design and construction inspection | Health and Safety Executive (HSE) | | | Health and Safety Executive NI (HSENI) |

| Topic | England | Wales | Scotland | Northern Ireland |
|---|---|---|---|---|
| Induced seismicity risk | British Geological Survey (BGS) | | | |
| Intention to drill, abstraction licensing, environmental permitting & Water Framework/Groundwater Directive objectives, fugitive emissions, environmental monitoring | Environment Agency | Natural Resources Wales | Scottish Environment Protection Agency | Northern Ireland Environment Agency |
| Hazardous substances / Expert advice on health impacts to environment agencies | Joint Agencies Groundwater Directive Advisory Group (JAGDAG) / Public Health England | Public Health Wales | Scottish Public Health Network | Public Health Agency Northern Ireland |
| Drinking water quality regulation | Drinking Water Inspectorate | | Drinking Water Quality Regulator Scotland | Drinking Water Inspectorate Northern Ireland |
| Water industry capital investment and water pricing | Water Services Regulatory Authority (OFWAT) | | Water Industry Commission for Scotland | Utility Regulator |
| Environmental effects | Department for Environment, Farming & Rural Affairs (Defra) | | Scottish Executive Environment & Rural Affairs Department (SEERAD) | Department of the Environment (DOE, Northern Ireland) |

groundwater monitoring requirements both in terms of establishing a baseline and during drilling/operation of the well(s). This contrasts with the attention given to induced seismicity and associated monitoring requirements.

The Environment Agency has gone some way to identify the environmental risks and has completed a high-level risk assessment.[37] It is also in the process of developing technical guidance to explain which environmental regulations apply to operations to explore for on-shore oil and gas in England and the permissions that need to be obtained. This will need to be developed further into more site-specific guidance for shale gas exploration/exploitation risk assessments to ensure adequate controls are implemented and risks managed effectively. The different source–pathway–receptor combinations that will need to be considered are illustrated in Figure 6.

The environmental regulators each have polices for the protection of groundwater.[38–40] These set out the legal (EU and UK) requirements to manage and protect groundwater (and the associated wider environment); the regulatory framework; environmental objectives; management and risk assessment tools/methodologies; and the supporting monitoring requirements.

The UK groundwater protection strategies adopt a risk-based approach in general but apply a precautionary approach in some instances where the consequences of impact may be so severe that the uncertainties associated with assessing risk are considered to be too great. An example of this is the inner source protection zone (or SPZ1) for public water supply abstractions. The precautionary principle is applied and certain activities (*e.g.* landfilling)

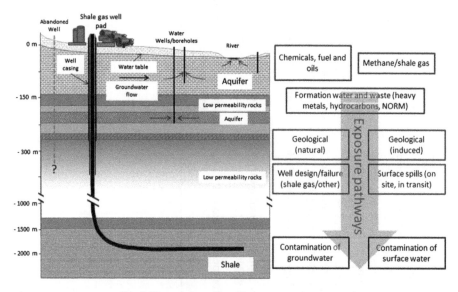

**Figure 6**  Conceptualised illustration of pollutants, pathways and receptors associated with shale gas operations.

are prohibited within SPZ1. The Environment Agency has indicated that it will also object to any shale gas infrastructure being located in SPZ1.[41]

In order for an operator to carry out an activity that may potentially impact groundwater (surface water or habitats) they must demonstrate to the environment agencies that they have assessed the risks to groundwater as part of the environmental permitting process. Whilst the environment agencies already have directly applicable procedures (as they permit many other industrial activities) there are unique aspects of shale gas exploitation that will need to be considered. This is particularly the case because the risks to groundwater are not only from activities on (or close to) the land surface – 'top down' – but also from below – 'bottom up'. This requires a new approach to groundwater vulnerability assessment and risk assessment. The British Geological Survey in partnership with the Environment Agency is currently mapping the 3D spatial relationship between shale gas source rocks and the principal aquifers of the UK as a preliminary step in this process (www.bgs.ac.uk/research/groundwater/shaleGas/aquifersAndShales).

## 7   Evidence of Shale-gas-related Groundwater Contamination

There have been only a few published peer-reviewed scientific studies that have assessed the impact of shale gas extraction on groundwater, but the number is increasing as concern has grown about the environmental impact of the industry. As the United States has seen the most rapid development of shale gas it is inevitable that the focus of most studies is here. One of the challenges that has emerged is the fact very little, if any, baseline monitoring took place before development, which has led to considerable uncertainty in attributing contamination directly to shale gas as the specific cause(s). In the UK, as development has not yet taken place, there is an opportunity to establish a pre-industry baseline and the British Geological Survey (BGS) has initiated such a study.[6] It is expected that the industry will also be required to undertake more localised baseline monitoring (as well as on-going monitoring) around any exploration/production sites as a condition of their environmental permit.

From the studies in the United States that have been published, the most common problems appear to be related to well integrity, where poor installation of wells and/or their degradation over time has been identified as a potential mechanism for contamination of shallow aquifers. As described earlier, studies of available datasets on well integrity show the extent of the problem.[46]

In 2007, a well that had been drilled almost 1200 m into a tight sand formation in Bainbridge, Ohio, was not properly completed and methane migrated upwards to contaminate a shallow aquifer and private water supply. A build-up of methane in the basement led to an explosion which alerted state officials to the problem.[42]

In aquifers overlying the Marcellus and Utica shale formations of northeastern Pennsylvania and upstate New York, Osborn *et al.*[43] have documented evidence for methane contamination of drinking water and attributed this to shale gas extraction. In active gas-extraction areas, average and maximum methane concentrations in drinking-water wells (19.2 and 64 mg $CH_4$ per litre, respectively) increased with proximity to the nearest gas well. At these concentrations there is a potential explosion risk if gas is allowed to accumulate in a confined space. In contrast, samples from groundwater in non-extraction areas (no gas wells within 1 km) within similar geologic formations and hydrogeological settings concentrations averaged only 1.1 mg $l^{-1}$. These results and conclusions were disputed by a number of other scientists on the basis that no baseline was available and it was known that groundwater in the area contained naturally elevated concentrations of methane. The study was extended by Jackson *et al.*,[5] with a larger number of samples and also with additional isotopic and geochemical analysis. This further work concluded that a sub-set of drinking water wells were contaminated as a result of drilling operations and that this was likely to be due to poor well completion (integrity).

Another well-publicised study at Pavillion, Wyoming,[44,45] investigated contamination of shallow groundwater supplies in an area with a considerable number of oil and gas wells (169) that had been hydraulically fractured. The study, which was led by the US Environment Protection Agency (USEPA), collected samples from both deep monitoring wells and shallower domestic wells. The study found that concentrations of dissolved methane in domestic wells generally increased the closer they were to gas production wells. An analysis of data on hydrocarbon well completion revealed a catalogue of problems with production wells having poor cement bonding or even no bonding present at all or over considerable lengths of the wells. Whilst these would almost certainly contribute to migration of contaminants to the shallow aquifers, other causes for concern and alternative pathways were identified. It was found that hydraulic fracturing and hydrocarbon exploitation took place at relatively shallow depths, from 372 m below ground level, whilst some drinking water abstraction wells in the overlying aquifer were as deep as 244 m. This means that there is only limited vertical separation between the two. A further problem identified was leakage and infiltration to shallow aquifers from storage pits on the surface used for storage/disposal of drilling wastes, produced-water and flowback fluids. The study is continuing but the USEPA have now, somewhat controversially, handed over responsibility to the State of Wyoming.

It is highly unlikely that such a situation would arise in the UK because of the strict controls that would be applied. However, the experiences from the United States serve as a clear indication of why, and what can happen if, things go wrong.

## 8  Conclusions

The UK may possess considerable reserves of shale gas with shale underlying a significant proportion of the UK, but as yet there has been very little exploratory drilling to confirm the resource potential.

In some parts of the UK the areas likely to be exploited for shale gas are overlain by significant aquifers used for drinking water supply and for supporting baseflow to rivers. The vulnerability of groundwater (and the wider water environment) to shale gas operations must therefore be taken very seriously.

Groundwater may potentially be contaminated by extraction of shale gas, both from the constituents of shale gas itself, from the hydraulic fracturing fluids, from flowback/produced water which may have a high content of saline formation water, or from drilling operations.

A rigorous assessment of the risks is required and appropriate risk management strategies need to be developed and implemented if the industry is to become established. It is likely that, due to environmental sensitivities, there will be some locations that shale gas exploitation will be considered unacceptable. To what extent this may affect the economic viability of the industry is unknown. In fact, there are very many unknowns as we are at the preliminary stages of exploration.

Because we are still at a very early stage we can take advantage of this and ensure that progress is made in a controlled way. We must identify and understand the risks to groundwater from shale gas and establish a fully informed risk management strategy for the industry. Experience from the United States has illustrated what can go wrong if this is ignored. We must not look back in 20–30 years and regret not taking the actions we have the opportunity to take now.

# References

1. I. J. Andrews, *The Carboniferous Bowland Shale Gas Study: Geology and Resource Estimation*, British Geological Survey for Department of Energy & Climate Change, London, UK, 2013.
2. R. C. Selley, *Geol. Soc. Petrol. Geol. London Conf. Series*, 2005, **6**, 707–714.
3. N. Smith, P. Turner and G. Williams, *Geol. Soc. Petrol. Geol. London Conf. Series*, 2010, 7, 1087–1098.
4. T. Harvey and J. Gray, *The Unconventional Hydrocarbon Resources of Britain's Onshore Basins – Shale Gas*, Department of Energy & Climate Change, 2010; www.og.decc.gov.uk/UKpromote/onshore_paper/UK_onshore_shalegas.pdf (last accessed 26/03/2014).
5. R. B. Jackson, A. Vengosh, T. H. Darrah, N. R. Warner, A. Down, R. J. Poreda, S. G. Osborn, K. Zhao and J. D. Karr, *Proc. Natl. Acad. Sci. U. S. A.*, 2013, **110**, 11250–11255.
6. British Geological Survey, *Baseline Methane Survey of UK Groundwaters*, 2013; http://www.bgs.ac.uk/research/groundwater/quality/methane_baseline_survey.html (last accessed 26/3/2014).
7. S. J. Mathers, R. L. Terrington, C. N. Waters and A. G . Leslie, GB3D: a framework for the bedrock geology of Great Britain, *Geosci. Data J.*, 2014; doi: 10.1002/gdj3.9.

8. British Geological Survey and Environment Agency, *Aquifer Designation Data*, 2013; http://www.bgs.ac.uk/products/hydrogeology/aquifer Designation.html (last accessed 26/3/2014).

9. M. E. Stuart and R. S. Ward, *The Potential Impacts on Groundwater from Shale Gas Exploitation in the UK*, Open Report OR/14/020, British Geological Survey, 2014; http://nora.nerc.ac.uk (last accessed 26/3/2014).

10. M. E. Mantell, Produced water reuse and recycling challenges and opportunities across major shale plays, *EPA Hydraulic Fracturing Study Technical Workshop #4 Water Resources Management*, EPA 600/R-11/048, 2011; www.epa.gov/hfstudy/09_Mantell_-_Reuse_508.pdf date (last accessed 26/3/2014).

11. Cuadrilla Resources, *Water Sourcing*, 2012; http://www.cuadrillare sources.com/protecting-our-environment/water/water-sourcing/ (last accessed 26/3/2014).

12. DECC, *Strategic Environmental Assessment for Further Onshore Oil and Gas Licensing. Environmental Report*, AMEC for Department of Energy & Climate Change, 2013. https://www.gov.uk/government/uploads/system/uploads/attachment_data/file/273997/DECC_SEA_Environmental_Report.pdf (last accessed 26/3/2014).

13. Environment Agency, *Managing Water Abstraction*, 2013; http://a0768b4a8a31e106d8b0-50dc802554eb38a24458b98ff72d550b.r19.cf3.rackcdn.com/LIT_4892_20f775.pdf (last accessed 26/3/2014).

14. E. Grubert and S. Kitasei, *How Energy Choices Affect Freshwater Supplies: A Comparison of U.S. Coal and Natural Gas*. Worldwatch Institute, Briefing Paper 2, 2010. http://www.worldwatch.org/system/files/BP2.pdf (last accessed 6/7/2014).

15. K. Johnson, K. French and J. K. Fichter, in *Corrosion 2008*, NACE International, New Orleans, 2008.

16. Cuadrilla Resources, *Fracturing fluid*, 2013; http://www.cuadrilla resources.com/what-we-do/hydraulic-fracturing/fracturing-fluid/ (last accessed 26/3/2014).

17. J. D. Arthur, B. Bohm and M. Layne, *Hydraulic Fracturing Considerations for Natural Gas Wells of the Marcellus Shale*, The Ground Water Protection Council 2008 Annual Forum, Cincinnati, Ohio, 2008.

18. D. M. Kargbo, R. G. Wilhelm and D. J. Campbell, *Environ. Sci. Tech.*, 2010, **44**, 5679–5684.

19. S. Basu, in *Workshop on Water Management in Marcellus Shale Gas Exploration and Production*, Atlantic City, 2011.

20. K. B. Gregory, R. D. Vidic and D. A. Dzombak, *ELEMENTS*, 2011, **7**, 181–186.

21. D. L. Shaffer, L. H. Arias Chavez, M. Ben-Sasson, S. Romero-Vargas Castrillón, N. Y. Yip and M. Elimelech, *Environ. Sci. Tech.*, 2013, **47**, 9569–9583.

22. B. D. Lutz, A. N. Lewis and M. W. Doyle, *Water Resour. Res.*, 2013, **49**, 647–656.

23. J. F. Barker and P. Fritz, *Can. J. Earth Sci.*, 1981, **18**, 1802–1816.

24. R. S. Ward, G. M. Williams and C. C. Hills, *Water Manag. Res.*, 1996, **14**, 243–261.
25. K. M. Révész, K. J. Breen, A. J. Baldassare and R. C. Burruss, *Appl. Geochem.*, 2010, **25**, 1845–1859.
26. T. Myers, *Ground Water*, 2012, **50**, 872–882.
27. J. E. Saiers and E. Barth, *Ground Water*, 2012, **50**, 826–828.
28. M. Zoback, S. Kitasei and B. Copithorne, *Addressing the Environmental Risks from Shale Gas Development*, Briefing Paper 1, Natural Gas and Sustainable Energy Initiative 2010, Worldwatch Institute.
29. New York State, *Supplemental Generic Environmental Impact Statement on the Oil, Gas and Solution Mining Regulatory Program*, New York State Department of Environmental Conservation Division of Mineral Resources, 2011; http://www.dec.ny.gov/energy/75370.html (last accessed 26/3/2014).
30. R. J. Davies, S. A. Mathias, J. Moss, S. Hustoft and L. Newport, *Marine Petrol. Geol.*, 2012, **37**, 1–6.
31. K. Fisher, *Am. Oil Gas Rep.*, 2010, July; http://www.aogr.com/magazine/frac-facts/data-confirm-safety-of-well-fracturing (last accessed 26/3/14).
32. Schlumberger, *Oilfield Rev.*, 2003, Autumn, 62–76.
33. CIWEM, *Shale Gas and Water. An Independent Review of Shale Gas Exploration and Exploitation in the UK with a Particular Focus on the Implications for the Water Environment*, Chartered Institute for Water and Environmental Management, London, 2013.
34. C. A. Green, P. Styles and B. J. Baptie, *Preese Hall Shale Gas Fracturing Review & Recommendations for Induced Seismic Mitigation*, Report to DECC, 2012.
35. DECC, *Oil and Gas: Onshore Exploration and Production*, 2014; https://www.gov.uk/oil-and-gas-onshore-exploration-and-production (last accessed 5/2/2014).
36. DECC, *Regulatory Roadmap: Onshore Oil and Gas Exploration in the UK Regulation and Best Practice*, 2013; https://www.gov.uk/government/publications/regulatory-roadmap-onshore-oil-and-gas-exploration-in-the-uk-regulation-and-best-practice (last accessed 26/3/2014).
37. Environment Agency, *An Environmental Risk Assessment for Shale Gas Exploratory Operations in England*, 2013; http://a0768b4a8a31e106d8b0-50dc802554eb38a24458b98ff72d550b.r19.cf3.rackcdn.com/LIT_8474_fbb1d4.pdf (last accessed 26/3/2014).
38. Scottish Environmental Protection Agency, *Groundwater Protection Policy for Scotland v3*, 2009.
39. Environment Agency, *Groundwater Protection: Principles and Practice (GP3)*, 2013; http://a0768b4a8a31e106d8b0-50dc802554eb38a24458b98ff72d550b.r19.cf3.rackcdn.com/LIT_7660_9a3742.pdf (last accessed 26/3/2014).
40. Northern Ireland Environment Agency, *Policy and Practice for the Protection of Groundwater in Northern Ireland*, 2001; http://www.doeni.gov.uk/niea/policy_and_practice_for_the_protection_of_groundwater_in_northern_ireland.pdf (last accessed 26/3/2014).

41. Environment Agency, *Onshore Oil and Gas Exploratory Operations: Draft Technical Guidance*, 2013; https://www.gov.uk/government/consultations/onshore-oil-and-gas-exploratory-operations-draft-technical-guidance (last accessed 26/3/2014).

42. Ohio Dept of Natural Resources, *Report on the Investigation of the Natural Gas Invasion of Aquifers in Bainbridge Township of Geauga County, Ohio*, Ohio Dept of Natural Resources, Division of Mineral Resources Management, 2008.

43. S. G. Osborn, A. Vengosh, N. R. Warner and R. B. Jackson, *Proc. Nat. Acad. Sci. U. S. A.*, 2011, **108**, 8172–8176.

44. D. C. DiGuilio, R. T. Wilkin, C. Miller and G. Oberley, *Investigation of Ground Water Contamination near Pavillion, Wyoming*, Draft Report EPA 600/R-00/000, US Environmental Protection Agency 2011; http://www2.epa.gov/region8/pavillion (last accessed 26/3/2014).

45. J. Tollefson, *Nature News*, 2012, 4 October, doi:10.1038/NATURE.2012.11543; http://www.nature.com/news/is-fracking-behind-contamination-in-wyoming-groundwater-1.11543 (last accessed 26/3/2014).

46. R. J. Davies, S. Almond, R. S. Ward, R. B. Jackson, C. Adams, F. Worrall, L. G. Herringshaw, J. G. Gluyas and M. A. Whitehead, *Mar. Pet.Geol.*, in press; http://dx.doi.org/10.1016/j.marpetgeo.2014.03.001 (last accessed 26/3/2014).

# Coal Seam Gas Recovery in Australia: Economic, Environmental and Policy Issues

ALAN RANDALL

ABSTRACT

Australia is experiencing a massive expansion of coal seam gas (CSG) extraction in response to buoyant international demand for liquefied natural gas and encouraged by accommodative mineral rights and taxation policies. The industry is capital-intensive and, while wages are high, employment of Australian workers is modest. The economic benefits accrue in the first few decades while the environmental costs may continue for a very long time.

The CSG and shale gas extraction processes are commonly quite different: for CSG, dewatering is the main method of releasing the gas, and fracking is at present used in only a minority of wells. In an arid land, dewatering raises major concerns of cumulative impact on groundwater systems, which can only be allayed by disconcertingly expensive wastewater treatment and recycling. Environmental impacts also include methane leakage into the atmosphere (which undercuts CSG's cleaner-burning advantage relative to coal), disturbance of sub-surface aquifers and geological structure, fragmentation of landscape, and disruption of ecosystems and agricultural production. Regulation of CSG extraction remains a work in progress, but is becoming more substantive in response to public concerns.

This chapter elaborates on the promises and challenges of massive CSG development, and discusses the relevant regulatory and taxation issues. Given that major Australian CSG developments lie beneath

Issues in Environmental Science and Technology, 39
Fracking
Edited by R.E. Hester and R.M. Harrison
© The Royal Society of Chemistry 2015
Published by the Royal Society of Chemistry, www.rsc.org

prime agricultural lands, I summarise the reasoning and empirical findings of a recent case study of the economics of competition and coexistence of CSG and agriculture on prime lands. Uncertainties and unknown unknowns are of such magnitude that they tend to dominate the policy discourse.

## 1   Introduction and Context

The massive growth of coal seam gas (CSG) extraction in Australia in the last 15 years, with projected future expansion dwarfing that experienced thus far, makes an interesting story in itself. Furthermore, the national and global context really matters. The Australian CSG sector is participating in a great global expansion of unconventional oil and gas extraction that, while promising inexpensive energy to fuel the on-going economic growth in many low- and middle-income countries, will exacerbate environmental threats that are serious already. The way these forces play out in Australia is framed by the particular structure of land and minerals rights that still reflect the country's British-colonial heritage and Australia's view of itself as resource-rich, capital-poor, and at the mercy of international markets and investors. Furthermore, and no surprise, the policy and regulatory debates around CSG invoke conflicting worldviews, and the conflicting fact-claims that arise when facts are treated as malleable in support of a particular worldview. I begin by elaborating a little on these three contextual matters, to set the stage for discussion of the economic, environmental and policy issues concerning CSG in Australia. This context-setting section is purposefully broad, impressionistic and documented sparsely. Facts and reasoning crucial to the subsequent discussion of CSG in Australia are documented more completely in subsequent sections.[1]

### 1.1   Global Energy and Greenhouse Gas Emissions Outlook

Recent analyses of the global energy outlook through 2030 project strong growth in energy demand, most of it driven by growth in emerging economies.[2,3] Much of the additional production to meet these demands will come from unconventional sources of oil and natural gas. Natural gas will be the fastest-growing fossil fuel, and liquefied natural gas (LNG) will be the most rapidly growing segment of the gas market. Nevertheless, the market share of fossil fuels as a group will diminish, even as the gas segment grows, while renewables make an increasing but still relatively small contribution.

Thoughtful people will greet these prospects with an ambivalence not easily resolved. Economic growth is to be welcomed, especially in emerging economies, and among well-off countries Australia and Canada are enjoying an energy export bonanza while the United States is anticipating an economic boom driven by cheap and plentiful domestic oil and gas. But even for these beneficiaries, the associated increase in projected global carbon emissions is a cause for serious concern, if not always for ameliorative policy.[4] The increasing market share of oil and gas provides some comfort – these energy sources, especially gas, burn cleaner than coal – but carbon emissions and methane leakage in extraction (especially by unconventional methods) and processing potentially offset this benefit in part or whole.

## 1.2   The Australian Context

Some particular Australian historical and institutional peculiarities shape and will continue to shape the way Australia experiences its boom in un-conventional gas extraction. The system of property rights reflects its British colonial history, but with a twist. In 1770 Captain James Cook claimed the east coast of Australia for Britain under the doctrine of *terra nullius* (nobody's land) thereby denying the legitimacy of all claims its original inhabitants may have had. It took until the 1990s before aboriginal land claims were given a rather limited form of recognition. Consistent with *terra nullius*, settlers established claims to land by establishing occupancy and only later were property rights for these squatters formalised. As this process wore on, 'public lands' became literally a residual category: lands held by government only because they remained unclaimed by settlers. Sub-surface rights, including mineral rights, remained with the Crown – in effect, the property of colonial governments, the state governments that succeeded them and the territorial governments set up in non-state territories following federation. Governments allocate exploration permits non-competitively and for nominal fees, transferring to the permit-holder the right to explore under relatively liberal conditions and to retain the rights to any minerals found – essentially a 'finders keepers' regime.

The net effect of this convoluted history is that surface and sub-surface rights are severed in most cases. Sub-surface rights are dominant and are readily appropriated by licensed explorers. In practice, this means that land-owners cannot look to property rights for defence against damages induced by mining; rather, their legal protections are limited by the will-ingness and capacity of governments to implement adequate regulatory regimes. Nor do governments retain property in minerals discovered. Vehicles for retaining benefits of mining for the Australian public are limited to royalties and taxation and, with roughly 83% of minerals activity controlled by multi-national corporations, governments have trodden lightly in these matters.

Australia's prosperity has long been underpinned by commodity exports, agricultural and mineral, in proportions that have varied according to cir-cumstances. Economic growth spurts were driven by mining in the 1850s through 1870s (gold), the 1870s–1900 (copper, tin, lead, and zinc), and the 1960s through the 1970s (iron ore, bauxite, nickel, tungsten, rutile, uranium, oil and natural gas). The current minerals boom, beginning in the early 2000s, has been driven by exuberant economic growth in Asia, especially mainland China. It focused initially on iron ore and coal, and was driven more by buoyant prices than by increased output. The current expectation is that the next leg of this boom will be driven by increased quantities of exports while prices are expected to moderate, and led by coal and LNG exports.

While the early minerals booms employed around 15% of workers in mining-related jobs, subsequent minerals booms have had substantial macroeconomic impacts on balance of payments, exchange rates, *etc.*, but mining-related employment has never recovered the prominence it enjoyed in the late 1800s. In 2012, a time of high minerals production and even greater investment in new mining and processing facilities, mining-related employment was little above 2%, but minerals accounted for almost two-thirds of the total value of exports.[5] Arguably, Australia's Reserve Bank, concerned to resist inflation and spread some of the minerals benefits to the broad mass of consumers during the current minerals boom, has also exacerbated the so-called 'Dutch disease' by maintaining high exchange rates that disadvantaged other exporting and import-substitution sectors.

This historical legacy has left Australia with property rights that encourage minerals exploration and extraction and favour the minerals sector over land-owners and communities, and a body politic that (for fear of killing the golden goose) is reluctant to charge adequate royalties and/or impose non-trivial mineral taxes that would ensure that Australia retains the economic rents attributable to the extraction of its exhaustible mineral resources.

## 1.3   Conflicting Worldviews

Unconventional fossil fuels pit market values against non-market values from the local to the global, *i.e.* from environmental concerns in the oil and gas fields and around the ports under construction to serve exports, to global climate issues. They pit short-term against longer-term considerations as the benefits from extraction accrue within 40 years while costs of land and water degradation continue long after the minerals operators have moved on. CSG extraction in Australia is concentrated in prime agricultural regions, pitting multinational minerals powerhouses against rural and traditional values on a playing field that seems tilted in favour of CSG. So it is unsurprising that CSG is controversial in many quarters.

Resolving these conflicts is not a simple matter of laying the facts on the table, because the facts themselves often are subject to dispute. Research has shown that worldview is cognitively prior to fact – *i.e.* worldview tends to shape what propositions people believe about the consequences of actions – so that mutually inconsistent views of particular prospects about, say, outcomes of CSG development may be manifestations of different worldviews.[6] While this insight has many interesting implications for conflict resolution, I raise it here only to suggest the difficulty of finding amicable resolution for many of the conflicts surrounding CSG development.

## 2   The Australian Minerals Economy

### 2.1   Exports

From less than half of Australia's exports (in value terms) in 1990, minerals exports had risen to almost two-thirds in 2012,[5] driven mostly by exuberant economic growth in Asia, especially mainland China, a bit-player among importers in 1997 but dominant by 2010. In the same period the proportion of energy minerals in the total mix increased only a little – while energy exports get a lot of press, iron ore exports also have experienced vigorous expansion. The composition of the energy component continues to change: thus far it has been dominated by coal, but LNG exports are substantial now and expected to grow exuberantly. Australia remains a net importer of oil.

Among the states, Western Australia still leads in total value of mineral production, with well-established iron ore exports and significant LNG projects on the horizon, but Queensland is gaining fast due to rapid expansion in the energy sector. Note that these are large states with population concentrated in a few urban centres, while minerals operations are mostly in remote locations.

### 2.2   Benefits to Australia

One can imagine a minerals boom benefitting the country in several ways – employment and wages, macroeconomic benefits and capital accumulation – but previous minerals booms have been followed by busts, and minerals extraction tends to be environmentally messy.

*2.2.1   Employment and Wages.*   The technology of mining has changed dramatically in the last century and the industry is now much more capital-intensive. Despite the on-going export boom, mining-related employment remained stuck at little above 2% in 2012.[5] Wages are relatively high, reflecting the high skill levels demanded and the need to attract workers to remote locations and challenging environments.

*2.2.2   Macroeconomic Benefits.*   Obviously, a major export boom helps a county's balance of trade and payments. The traditional Australian way of sharing the benefits of export booms among the broad population is to allow exchange rates to rise which, as well as moderating potential inflation, serves to make imported consumer goods and international vacations more affordable for the broad mass of consumers. Nevertheless, there are persistent complaints that while the minerals sector is booming the rest of the economy just plods along, a phenomenon often described as the 'two-speed' economy. If we introduce the 'Dutch disease' argument – that the high-exchange-rate solution disadvantages other exporting and import-substitution sectors – a case can be made that it is really a 'three-speed' economy: the thriving minerals sector, the rest of the tradables sector which is depressed to varying degrees, and the plodding non-tradable sector.[7] The project of sharing the benefits of the minerals boom broadly has thus far achieved very incomplete success.

It now seems clear that the current minerals boom comes in two stages. The initial stage has been driven more by high minerals prices than increased output. Observed productivity in the industry has been declining, which might have a couple of explanations: willingness to use more inputs and to work lower quality deposits in direct response to the higher prices; and the price-driven export boom has led to an investment boom in the industry, with inherently high up-front costs. The second stage will be driven more by increased quantities of exports, as exploration and investment in the initial stage comes to fruition.[8] We may well expect the stage one productivity decline to be reversed in due course.

*2.2.3   Capital Accumulation.*   Minerals are exhaustible resources. Extraction means depletion, as opposed to the renewable resource case where harvest may be sustainable if the resource is managed carefully. But all is not lost. Solow[9] and Hartwick[10] have shown that, under ideal conditions, an economy that relies on exhaustible resources may nevertheless achieve weak sustainability provided the economic net benefits (technically, the resource rents) from resource depletion are invested in reproducible capital. This kind of sustainability is called 'weak' because it sustains welfare, *i.e.* standard of living, while tolerating liberal substitutions in consumption and production. In contrast, strong sustainability requires that the depleted exhaustible resources be replaced directly with equivalent renewable resources.

The challenge facing a nation that exports exhaustible resources is to capture the resource rents and to ensure that they are, in fact, reinvested efficiently. Note that one would expect profits in the minerals industry to come in two parts – rewards to effort in extraction and processing, and resource rents – and it is not always easy to separate them empirically.

The actual and potential profits from the minerals industry are huge but, because the industry is predominantly foreign-owned – consensus estimates suggest at least 83% foreign ownership in the minerals sector[8] – much of the profits will leave Australia.[11] To meet the Solow–Hartwick sustainability criterion, Australia would need to intercept and retain the resource rents accruing initially to the minerals operators. Instruments for this purpose include royalties and taxes of various kinds, including severance taxes on extracted resources and taxes on mineral company profits.

While the industry will pay substantial royalties to state governments – in 2007 (the most recent year with complete statistics) the industry paid AUD6573 million in royalties on a gross output value of AUD106 216 million[5] – it remains an open question whether royalties and taxes on the industry are high enough to compensate Australians for the eventual exhaustion of a valuable resource. With much of the profits from extraction shifted offshore, the magnitude of revenue collected *via* royalties and taxes matters crucially to the nation. Despite the dominant position of Australian governments as sub-surface rights-holders, analysts conclude that they have left a considerable proportion of the net benefits on the table, *i.e.* failed to collect a substantial portion of the economic value of the nation's mineral resources depleted.[12]

*2.2.4   Boom and Bust.*   The projected economic benefits of CSG are not assured in the long run. While current projections are for high and stable commodities prices for the life of the planned projects, the extractive industries historically have experienced cycles of boom and bust.[13,14] In Australia there is an on-going debate as to whether the present boom also will exhaust itself in due course. The optimistic case depends on a number of strong assumptions: the economic momentum of the major Asian importing countries maintains a smooth upward path; Australia, despite its high cost structure, will continue to withstand competition, especially from shale gas extracted in other countries; and so on. The situation regarding competition among exporters may be volatile: substantial price differentials persist in international LNG markets, effectively insulating Australian exporters from competition from much cheaper North American shale gas.[15] Economists would expect these price differentials to be diminished or eliminated as time passes.

*2.2.5   Damage, as well as Benefit.*   The economic concept of benefit is directed ultimately to the bottom line. There is a conceptual symmetry of benefit and cost: a cost avoided is a benefit and a benefit foregone is a cost. This sounds complicated, but clarity readily is regained if we place benefits and costs on the same scale, assign positive (negative) value to things that make us better (worse) off, and sum it all to calculate net benefits. From that perspective, a consideration of benefits also must consider the costs of any damage resulting from extraction, processing, and transportation of minerals. All of these processes are environmentally intrusive to some degree, and many of them impose damage, not always fully compensated, on surface-based private and public activities and on groundwater. The costs of consequent damage are not always easy to measure, but it is well-established that they can be large.

## 2.3   Australian Minerals Rights and Resource Taxation Policies Encourage Extraction

It is well-known that minerals rights and taxation regimes influence the rates of extraction and the distribution of rewards therefrom.[16,17] In addition to generating revenue for government and providing an instrument for managing the rate of extraction, it was argued above that minerals taxes might be used to generate capital reserves for reinvestment to promote economic sustainability. Thus minerals rights and taxation policies are important considerations in any discussion of CSG mining and the national interest.

In Australia, sub-surface rights are separated from surface rights and re-tained by the Crown.[18] Surface rights come in several forms, predominantly long-term leases from governments, and freehold title. Sub-surface rights are dominant over surface rights in the sense that protections for surface-rights holders who may be impacted by sub-surface extraction are limited to those provided explicitly by statute law and regulations.

Traditionally, federal, state and territory governments, the sub-surface rights holders in Australia, allocate exploration and production rights to private investors and collect a return for the public *via* a mix of arrange-ments, predominantly royalties and taxes.[12] Commentators have described Australia's rights regime for minerals as effectively 'finders keepers'.[19,20] Exploration licenses are issued inexpensively and non-competitively, and license-holders are encouraged to explore actively. Licensed explorers who find potentially profitable deposits are awarded extraction leases, so that discoveries belong effectively to the finder. The states collect royalties, typically 10% *ad valorem* at the well-head for CSG,[21] which likely is well short of the full economic value of the nation's mineral resources depleted,[12] a stance that is tilted toward rewarding extraction excessively. Researchers have concluded that such regimes encourage extraction.[20,22]

## 3   Coal Seam Gas Development in Australia

There is a lot of economic momentum behind the expansion of coal seam gas (CSG) mining in Australia, driven by buoyant international demand for liquefied natural gas and an accommodative minerals rights and taxation regime. From a trivially small baseline in 1995, CSG is projected to provide about one-half of Australia's total gas output by the mid-2020s. Queensland is the state that led the way in terms of projects operating and committed, CSG production and CSG reserves remaining (see Figure 1). By 2012, annual production of CSG was 252 petajoules (PJ) in Queensland and 6 PJ in NSW, accounting for around 35% of Australian east coast gas consumption.[23] Queensland was projected in 2010 to have about 40 000 wells producing CSG by 2030.[24] Ongoing exploration may add to that number, but a modest contraction in export projections, reflecting increased supply from com-peting exporters, may have the opposite effect. Much of the output will be exported in several forms, the most prominent being liquefied natural gas (LNG) with projected exports of 16 million tonnes (Mt) by 2015.[24] New li-quefaction facilities and export terminals are under development on the east coast of Australia, to enable export of CSG. To meet known domestic and overseas commitments, including new LNG projects, the rate of drilling CSG wells in Queensland is forecast to intensify during 2014–15.[25]

Even at a relatively high rate of development, Australia is thought to have about 100 years of CSG reserves.[24] The largest reserves of CSG are in Queensland's Surat and Bowen basins, while the CSG reserves in NSW are relatively small (see Figure 1). Conventional gas remains an important re-source, with substantial offshore reserves near Victoria and terrestrial

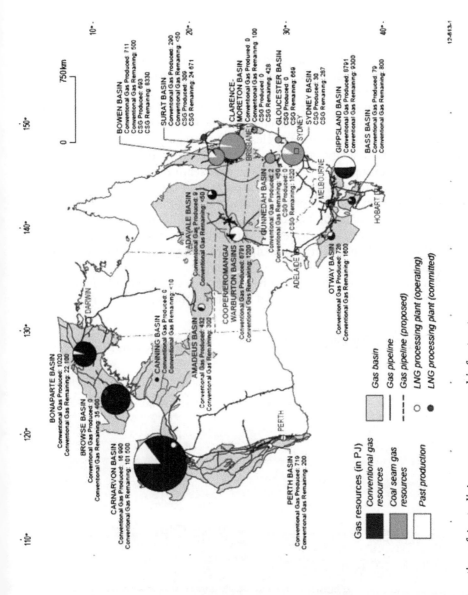

**Figure 1**   Location of Australia's gas resources and infrastructure.
(Source: Bureau of Resources and Energy Economics, *Australian Gas Resource Assessment*, 2012, Canberra).

reserves in the Cooper Basin in eastern Australia. Western Australia has very large conventional gas reserves offshore, supplying local demand and providing LNG exports. Shale gas exploration in Australia is still in its infancy but recent reports suggest enormous reserves and a potential for shale gas extraction eventually to dwarf CSG.[26]

## 3.1  CSG Extraction Technology

Coal seam gas is found in cracks, pores and micropores in coal seams, where it is held in place either as free gas or adsorbed onto coal surfaces. It is important to point out that the extraction technologies for CSG and shale gas are commonly quite different. Fracking – hydraulic fracturing, in which water mixed with sand and various chemicals is forced at very high pressure into shale beds or coal seams to release trapped gas – is fundamental to shale gas extraction, whereas the fundamental process for CSG extraction is dewatering, *i.e.* pumping water out of the coal seam to release trapped gas. The 'produced' water and the gas are separated at the surface and each is removed in its own network of pipes (see Figure 2), the gas to processing facilities and eventually to market, and the water to holding ponds and eventually (it is hoped) to treatment and recycling. In most cases the CSG is naturally of 'pipeline quality' and, apart from drying, requires minimal treatment.[27]

As stated above, fracking is not universal in CSG extraction, but neither is it entirely absent. A rule of thumb used in Queensland is that fracking occurs

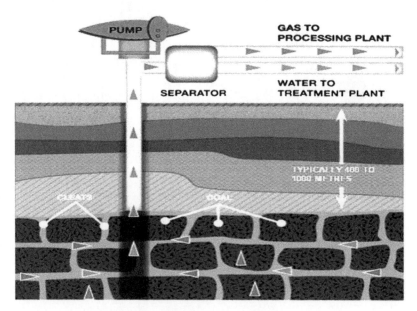

**Figure 2**   The CSG extraction process.
(Source: Natural CSG, http://www.naturalcsg.com.au/ last accessed 28/01/2014).

**Figure 3**   CSG landscape, QLD.
(Source: Stop CSG, http://stopcsg.org/csg-risks/ last accessed 28/01/2014).

in about 10% of new wells but may eventually be used in as many as 40% of aging wells in order to release the remaining but less accessible gas.[28] In addition, fracking may become more prevalent as the most attractive reserves are fully developed and CSG extraction moves on to less favourable sites. Seams that have plenty of natural fractures are less costly to develop, and are thus more attractive to CSG operators, than those that are more solid and need to be artificially fractured.[26] Most current CSG extraction in Australia is occurring 250–1000 metres below the land surface. However, as Australian production taps into deeper coal seams or those less naturally permeable, the need for fracking may increase from the current 10% of new wells to upwards of 40%.[29]

The issues surrounding fracking chemicals that are familiar to observers of the US scene – reluctance of operators to provide information about what chemicals and substances are being used, and instances where toxic and/or carcinogenic chemicals have been identified in the vicinity of fracking sites – also have arisen in Australia. CSG operators in Australia now are required to announce what substances are used during fracking, an obligation that seems to have motivated increasing use of environmentally friendly chemicals.[30]

CSG wells are typically networked across the landscape 750 m apart, linked by service roads and by at least two sets of pipelines to move gas and waste water out, and potentially a third set to bring fracking water in (see Figure 3). Waste water is collected in holding ponds from where the announced intent, currently honoured more often in the breach, is to move it on to treatment plants.

## 3.2   Environmental Impacts of CSG

CSG extraction on a large scale is a land-extensive but nevertheless intrusive process entailing a considerable catalogue of potential environmental risks and land-use conflicts – diminished water supply and quality, methane leakage into the atmosphere, disturbance of sub-surface aquifers and geo-logical structure, fragmentation of landscape and disruption of agricultural production.[1,24,30-34] The magnitudes of these threats are not merely un-certain in the statistical sense; in some cases they are driven by complex systems that work in ways we do not fully understand, even conceptually. In the face of these uncertainties and unknowns, the above-mentioned separation of surface and sub-surface rights limits the protections for landowners. Their protection, like that of the general public, depends on the willingness and capacity of governments to implement adequate regulatory regimes.

*3.2.1  Water Issues.*  CSG extraction on the scale planned in Queensland 'produces' vast quantities of waste water contaminated by chemicals found in and around the coal seams and sometimes introduced in the fracking water. Because aquifer systems are multi-layered and inter-con-nected, and groundwater and surface water are components of a single hydro-system, dewatering has system-wide impacts.[1,23,33,34] Contaminated waste water is 'produced' in vast quantities:

- Groundwater is depleted and water tables fall, with volume and quality implications for other users and groundwater-dependent ecosystems, and unwanted surface water interactions with groundwater occur over the short and long term;
- Multiple projects in an inter-connected system will have cumulative impacts;
- Water from surface sources and other aquifers will eventually flow into the depressurised aquifers around the coal seams to establish a new hydraulic equilibrium;
- Altered hydraulic gradients may produce mixing and cross-contamination between different aquifers and between groundwater and surface waters with different quality characteristics;
- Gas may migrate into surrounding aquifers, wells and water bores, and surface waters;
- Reduced water pressure in sub-surface layers may lead to compression of layers, alteration of hydraulic properties and subsidence at the surface;
- Surface and groundwater will be substantially less available in the new equilibrium unless waste water is recycled effectively; and
- Two kinds of recycling can be conceived: treated water is re-introduced to surface steams and alluvial aquifers, and waste water (with minimal treatment) is re-injected into the coal seam. Both would tend to re-equilibrate the hydraulic system, but in very different ways and with very different economic consequences.

This vast and complex system does not lend itself to compartmental-isation, but a little bit of *ad hoc* labeling and grouping of impact categories may help the exposition.

### 3.2.1.1 Groundwater Depletion

'Produced water' that is not recycled or re-injected depletes aquifers; in the fullness of time, the vacuums created will be filled with water seeping in from elsewhere, lowering water tables in general. The quantities of withdrawals for CSG extraction will be truly vast. Because so much of the CSG action will be concentrated in the Great Artesian basin (GAB), basin-level analysis is essential. According to the National Water Commission (NWC),[34] planned CSG development will, at full operation, withdraw more than 300 gigalitres of groundwater annually from the GAB, *i.e.* more than 60% of total allowable withdrawals. This 60% for CSG implies some combination of displacing existing uses and pushing total withdrawals well above sustainable levels. The NWC estimate is thought to be relatively conservative: CSG industry sources offer a somewhat lower projection, but the federal government's 'Water Group' suggests, based on its case studies of the Surat and Bowen sub-basins, that GAB-wide withdrawals may exceed the NWC estimate considerably.[35]

The theory of complex systems suggests that it is near-impossible to predict the cumulative impacts on groundwater over several centuries, because the GAB hydrological system is much too complex and the cumulative shock to the system from CSG development will be much too large to be characterised with standard groundwater models and modeling methods.[31,36] The Queensland Water Commission (QWC) has made a serious effort to build a model to analyse the cumulative impact on groundwater of planned CSG development in the Surat region.[37] They model movement of water among twenty geological layers from the surface down well below the coal seam, in response to planned dewatering. The models are best described as complicated rather than complex, because they are essentially reductive and therefore unable to capture the emergent possibilities in complex systems. Nevertheless, the effort is to be applauded because it is really important to make progress in projecting cumulative impacts and we should not be too critical of reductive modeling approaches, given the relatively tentative progress that has been made in complex systems modeling. QWC predicts serious depletion of water tables in some parts of the Surat, and the specific bores and wells most likely to be impacted have been identified. Intuition suggests that their projections may understate the eventual impacts for two reasons: unaccounted complexity of the system is likely to magnify impacts, and impacts are likely to persist and perhaps accelerate beyond the time horizon of their models.

At the site and project levels, farmers and settlements using artesian water worry that water pressures and levels will fall, wells and bores will need to be drilled deeper, and some may dry-up completely (and the QWC models validate that concern in some cases). The Queensland government has insisted on 'make good' provisions, requiring operators to provide water to land users facing reduced and more expensive groundwater supplies as a result of CSG activity.[38] 'Making good' is intended to compensate in-kind for any harm that may arise from extracting water to release the gas, but three kinds of operational difficulties are obvious: establishing the cause and effect relationship with CSG extraction; the increasing infeasibility of making good as the cumulative impacts of CSG extraction grow large with the increasing number of wells across the landscape; and reconciling the long-term impacts on aquifers, which are likely to play out on a time-scale of many decades and perhaps centuries, with the much shorter time-scale of CSG extraction.

There are reports from the Surat that Arrow Energy, a major CSG operator, is delivering treated water in tanker trucks to farmers who have complained of CSG-induced water shortages.[39] This is economically feasible for a small number of impacted livestock operations, but it also illustrates the limits of making good inexpensively: this solution will become less feasible as more farmers are affected and the travel distances become longer, and is obviously infeasible if water for broad-acre irrigation is impacted.

### 3.2.1.2   Groundwater Contamination

Because CSG wells penetrate many geological layers and aquifers in order to reach the coal seams, deterioration of the cement casing that lines the well could allow cross-layer movements of contaminants.[40] Ensuring well integrity is therefore an essential element in managing potential contamination of groundwater resources as a consequence of CSG operations. Stricter standards for well preparation have improved well integrity in the last decade.[27] Nevertheless, the potential impacts of CSG wells on groundwater extend for centuries, while the economic life of CSG wells lasts for only a few decades at best, a situation that calls for rigorous inspection of wells, including those no longer in use, and mitigation of any problems revealed by inspection. Ideally such a program would be funded from sinking funds established by the CSG operators, but no such arrangement exists currently. This idea could be applied to environmental damage more broadly: a serious case can be made for requiring CSG operators to post environmental bonds consistent with the worst-case damage scenarios.[41]

### 3.2.1.3   Waste Water

The water co-produced in CSG extraction is briny to varying degrees and contains a range of chemicals naturally present in and around the coal seam. Depending on site conditions, these may include metals and radionuclides that can be toxic to plants, animals and humans.[29,33,34,42,43] The process of fracking, where used, may add to the chemicals in waste water. While the industry insists it is not presently using them, BTEX (benzene and similar organic chemicals thought carcinogenic) chemicals have in the past been added to the water. Volumes of 'produced' groundwater are typically large in the early stages of CSG production, and the volumes of gas released are small. However, later in the life of a well (which can be several decades) the water produced decreases and methane production increases. Seams that need fracking may produce less water than other seams.[27]

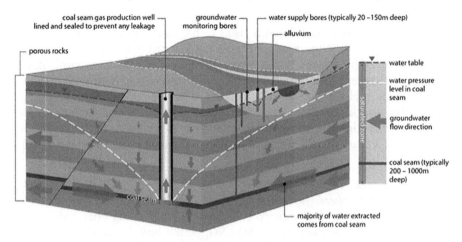

**Figure 4**  Groundwater flows in vicinity of CSG extraction well.
(Source: NSW Office of Water, http://www.water.nsw.gov.au/Water-man-
agement/Groundwater/Water-and-coal-seam-gas/Water-and-coal-seam-gas
last accessed 28/01/2014).

Contaminated waste water needs careful storage, transportation, and
treatment. Damage arises when it is spilt or leaked into crops, native vegetation, surface waters or shallow and deeper groundwater, which are connected components of the one hydrological system (see Figure 4).

Recycling the waste water represents the only conceivable way to compensate for the huge volume of groundwater to be extracted by the industry.
Treatment and recycling are processes that separate the waste water into two
components, a treated/recycled component that, depending on the level of
treatment, is safe for certain uses, and a solid and/or liquid component
(sludge) with concentrated salt and chemical contaminants that requires
safe disposal. The simplest and cheapest treatment methods, evaporation
ponds, contribute nothing to recycling, whereas reverse osmosis (basically,
desalination) is very expensive.[44]

In addition to the costs of recycling plants, recycling at scale requires an
extensive network of pipes to bring waste water from spatially dispersed sites
to a central recycling plant. If recycling waste water becomes the norm, recycled water will be produced in such volumes that the environmental impacts of returning it to the environment will raise issues: for example,
releasing recycled water into surface streams may produce sustained flows
where ecosystems demand episodic flows; and successfully re-injecting it
into depressurised aquifers in order to recharge the groundwater may over-
tax our technical capacity and our understanding of complex aquifer systems
in the coalfields.[32] Re-injection to restore the hydraulic pressure in aquifers
that have been over-pumped is seen as a worthy long-term goal, although it
appears to be a non-trivial process in terms of engineering feasibility and
cost.[23] Alternative solutions, including deep-ocean burial, have been can-
vassed but thus far have failed the feasibility test.

Arrow Energy is operating a reverse-osmosis plant of modest capacity in the Surat, but this is still the exception rather than the rule. Under current regulations, evaporation ponds are no longer permitted, but holding ponds of similar design provide 'temporary' storage while operators await development of cost-effective recycling methods. Regardless of treatment method, by-products include salt in vast volumes and contaminated sludge in quantities and kinds that depend on local conditions and extraction and treatment practices. Until better solutions are discovered, most of the contaminated waste will be stored in brine ponds and salt pits on the gas fields.[45]

*3.2.2    The Atmosphere.* CSG burns much cleaner than coal: typical carbon emissions per unit of electricity generated from burning coal range from 43% to 87% greater than from CSG-LNG, which has the additional advantage of not producing by-products such as sulfur, mercury, ash and particulates.[46,47] If emissions were restricted to those from burning fuel, wide-spread substitution of CSG for coal would bring big reductions in Australia's carbon footprint. More comprehensive accounts suggest a more nuanced picture – when the methane and carbon dioxide that will inevitably escape from CSG wells and gas-fields, the energy used and emissions released in extraction and processing, *etc.*, are counted, the greenhouse gas reduction benefit of CSG becomes more tenuous and specific to individual CSG operations.[48]

The sheer magnitude of planned CSG operations in Australia generates some very large estimates of potential atmospheric emissions. Approved CSG and liquefied natural gas projects, including associated infrastructure, could generate 39 Mt of carbon dioxide equivalent each year.[24] Modeling suggests that the CSG industry eventually could produce as much greenhouse gas as all the cars on the road in Australia.[24]

The cognisant federal agency has recognised the policy and regulatory challenge posed by large quantities of methane and carbon gases likely to be released directly into the atmosphere from the gas fields and liquefaction plants. Greenhouse gas data for CSG are being collected, including the primary sources of emissions and reasons for variations in leakage rates.[49] Well integrity is one component of a more comprehensive strategy to minimise leakage of CSG into the air.[50]

*3.2.3    Sub-surface Impacts.* Sub-surface geology and hydrology can be disturbed by CSG mining in two distinct ways: withdrawal of large quantities of water, which is endemic to CSG operations; and fracking, which fractures the coal seam and surrounding soil and rock layers to release the gas, and is used in some CSG operations. Gas extraction and the lowering of water tables create voids that may lead to land subsidence. Fracking may lead to disturbance and irreparable damage to aquifers, migration of methane and contaminants, and increased seismic activity.[30] The potential for these impacts is well recognised, but the extent of their realisation in Australia's CSG fields will emerge only with the passage of time.

*3.2.4   Fragmentation of the Landscape: Ecosystems and Agriculture.*   By its scale and nature, the 'footprint' of a CSG extraction field of this type cuts across landscape and biological habitat (see Figure 3). CSG extraction is a spatially dispersed industry with a much greater footprint on landscape and environment than the fairly modest surface area devoted to well-heads would suggest. As noted above, CSG wells typically are scattered across the landscape at 750 m intervals, but new horizontal drilling technology may allow well pads to be spaced farther apart. The networks of pipes for fracking water, gas and waste water, along with the processing, waste storage and treatment facilities, the network of roads needed to tend the wells and transportation up-grades to get the CSG products to market all contribute to landscape fragmentation, with negative impacts on agriculture and ecosystems.

In preparation for installing this CSG infrastructure, vegetation is removed. As with any activity that requires land clearing, this fragments the habitat and breeding ranges of native fauna such as lizards and birds, and can lead to the introduction of invasive species. The negative impacts of fragmentation of bushland on native fauna are well-documented.[51-53] Where a landscape already has been cleared extensively for urbanisation or agriculture, the remaining vegetation is often of high ecological value.[52,54] Regarding ecosystem impacts, the Australian Broadcasting Commission reports instances in which the federal government has granted major CSG operators permission to clear land containing species and ecosystems protected under federal threatened species legislation.[24]

Landscape fragmentation associated with CSG development in farming regions reduces agricultural productivity and increases farmers' costs, as do the groundwater depletion and resource contamination impacts of CSG operations.[1,27] Despite the major operators' commitment to good neighbour policies, the dominance of sub-surface rights disadvantages surface-rights holders by weakening their bargaining position. In regions with active or potential CSG operations, organisations have arisen to express concerns about the potential impacts of a weakened agriculture on rural and community ways of life.[55] Conflicts between agriculture and CSG strike with particular force in some of Australia's most productive farming areas, including the Darling Downs and the Liverpool Plains, where the national interest in prime farmland (quite scarce in Australia) comes into play, in addition to local concerns. Cases have been reported where the local Catchment Management Authority has developed integrative and cumulative impact assessment and risk management processes that smooth the path to coexistence of CSG and agriculture,[27] although coexistence agriculture is usually less productive and generates lower economic rents than agriculture-only.[1]

Some would argue that there is a national interest in preserving the very best farmland for agriculture, even if the economic argument for CSG development is strong.[56] However, the possibility might be considered that, for the best land and with a full economic accounting of costs as well as benefits, the economic advantage of CSG is not so compelling. CSG offers the prospect of several decades of lucrative extraction but it is reasonable to expect environmental costs, some of them potentially substantial in cumulative effect, to continue perhaps for long after the gas is gone.[1] Furthermore, the economic benefits of CSG are not assured: while current projections are for high and stable commodities prices for the life of the planned projects, the extractive industries historically have experienced cycles of boom and bust.[13,14] At best CSG is a transition energy technology and we do not know how long its window of opportunity will be.

*3.2.5 Coastal and Marine Environments.*   There is a massive on-going expansion of port and terminal capacity along the Queensland coast, purpose-built to serve huge coal transport ships and mega-tankers for LNG. Existing coal ports near Gladstone are being expanded and a new LNG facility is under construction. Four planned coal ports would extend almost 1000 miles north to Wongai, far up the Cape York Peninsula. The impacted coastal strip parallels the Great Barrier Reef, a World Heritage site that parallels the coastline typically fewer than fifty miles offshore.

Already the Reef is under pressure from coral bleaching and is threatened by climate warming. In 2011 a bund (*i.e.* retaining) wall in the Gladstone port leaked, leaching dangerous chemicals into the harbour and was suspected of causing morbidity and mortality of fish, turtles and dugong, and a sharp drop in the local fishing catch. The cause of the breach is disputed – the Ports Corporation blames heavy rain and floods – but a reasonable inference is that the millions of tonnes of dredging and dumping undertaken made the wall susceptible to the intense rain and flooding that occurred.[57]

Major environmental issues include the volume of dredging and dumping that will accompany the proposed expansion of facilities and the impact of greatly increased traffic of mega-ships in the narrow and shallow waterway separating the Reef from the shoreline.

The UNESCO World Heritage Center is considering placing the Reef on the List of World Heritage in Danger. The relevant committee's recent resolution requested that Australia "...not permit any new port development or associated infrastructure outside of the existing and long-established major port areas within or adjoining the (Great Barrier Reef)...".[58] Planned projects that would be affected include the Balaclava Island Coal Export Terminal, the Fitzroy Terminal Project and the Wongai Project, which are arguably the most damaging of the proposed expansions due to their location in largely undeveloped or near-pristine areas.[57]

*3.2.6 CSG and Rural Communities.*   CSG operations have a variety of impacts on nearby rural communities. The development phase employs many more workers than the subsequent operations phase, so communities must cope with a relatively sudden increase in demands for private and public sector services, followed after several years by withdrawal of a substantial portion of these demands. The need to ramp-up community infrastructure stresses the capacity of local government and the social fabric of the community. Communities that have viewed their prosperity as dependent on a thriving agriculture may be concerned that CSG operations threaten agricultural viability. Among the residents of impacted communities there will be some who are concerned about environmental as well as infrastructural impacts, and some who perceive a loss of community self-determination in the face of an economically powerful industry.

Given the intensity of planned CSG development in Queensland over the next few years, particularly in the Surat Basin, extraordinary demands will be made on rural infrastructure, housing and community services in health and education. As a result, communities are likely to raise concerns about the adequacy of infrastructure.

The dominant position of sub-surface rights provides few protections to land holders. With conventional open-cut (*i.e.* surface) coal mining, mine operators typically purchase the land at valuations well above commercial value, a practice that tends to minimise disputes with property holders about access. CSG operations are extensive and spatially distributed, discouraging total acquisition of impacted land-holdings. Despite access agreements, which compensate land owners for direct costs imposed by mining operations (with the amount of compensation determined by mediation in the event of disagreement), and good neighbour agreements voluntarily offered by CSG operators, some land holders remain resistant to co-existence with CSG operations. Communities have at least two kinds of stake in these conflicts: some residents will sympathise with land holders resisting co-existence, and resolution of conflicts between surface and sub-surface rights holders will not by itself assuage community concerns so long as there are spill-over effects on the community. Community resistance has led to the abandonment of at least one CSG development proposal in NSW (albeit by a relatively marginal operator in the industry).[59]

Citizens in several residential communities have been resisting co-existence with CSG industry development in Queensland and NSW.[38,60,61] The New South Wales government, in response to rising public concern, recently restricted CSG operations within two kilometres of residential areas or industry cluster areas.[62]

Patterns of social and economic impact on communities appear to depend on the size of the CSG project, community structure and history, and the extent to which a non-resident work force is involved. The level of local support for resource development including CSG is contingent upon economic benefits and opportunities accruing at the community level.[63] The local or regional economic multiplier effect of expenditures by CSG operators depends crucially on the extent of economic leakage, *i.e.* the proportion of operators' and workers' expenditures that impact outside of the immediate region. Smaller and less economically diverse regions tend to experience greater leakage.

Community structure and resilience are important to the ability of communities to cope with the sudden change unleashed by rapid CSG development. Researchers have identified five dimensions particularly important for community group resilience: strategic thinking, links within communities, effective use of resources, commitment, and building meaningful relationships.[64] These dimensions, and the qualities underpinning them, also contribute to resilience of the wider community. A diversity of groups, groups acting as bridging organisations and groups involved at different scales, all provide resilience to the wider system.

The Queensland government requires a social impact management plan for new developments,[65] and the notion of community and its complex social dynamics should be more carefully considered in social impact assessment.[66] CSG developers are being asked to accept an increasing set of responsibilities for community health, safety and social well-being. There is some evidence that operators are rising to the challenge: Santos, for example, has announced good progress in new housing construction in southern Queensland.[67]

*3.2.7   Toward a National Regulatory Framework.*   Regulation of CSG development and operations remains a work in progress.[38] As is typical in Australia, where regulation is mostly a state responsibility, regulation of CSG development and operations proceeded initially without interstate coordination, and early efforts were addressed more to process than substance. Ironically, overlapping and repetitious regulatory processes allowed Queensland CSG operators to claim that they were the most thoroughly regulated on earth, while substantive provisions were, on balance, rather accommodative.[38] Yet regulatory stances have become stronger in recent years – examples include Queensland's recent commitment to require assessment of the cumulative impacts of multiple projects on surface and groundwater, and New South Wales' recent adoption of a two kilometre set-back from residential areas and industrial clusters – as it has become increasingly obvious that land-holder and community opposition to CSG is not a passing fad.

The Commonwealth has stepped in, not as a regulator, but as a convenor of governments to offer a *National Harmonised Regulatory Framework for Natural Gas from Coal Seams* to serve as guidance regarding best regulatory practice for governments, industry and communities.[23] It details approaches agreed between the Australian state and federal Ministers responsible for resources, to provide guidance on leading practices for CSG operations, based on state and federal policies, legislation and regulations. Particular attention is paid to well integrity, water management and monitoring, fracking and management of chemicals. The time horizons disconnect between the long-term impacts on the environment broadly defined and the relatively short-term flow of revenues from CSG comes into full focus when addressing decommissioning and well abandonment. The National Harmonised Framework, if adopted by the states and territories (and the usual outcome of this kind of effort in

Australia falls short of universal adoption), would require that decommissioning and well abandonment ensure the environmentally sound and safe isolation of the well for the long term; the protection of groundwater resources; isolation of the productive formations from other formations; and the proper removal of surface equipment.[23]

## 4   Toward a Comprehensive Economic Assessment of CSG Development: a Case Study of CSG *versus* Agriculture on Prime Farmland

The foregoing discussion has established that CSG development in Australia is proceeding apace, driven by buoyant export demand but facilitated also by accommodative mineral rights and taxation regimes, the subordinate rights position of land holders, and regulatory structures that may be over-matched in their quest to internalise the industry's environmental impacts and reconcile its large but relatively short-term profits with its continuing adverse effects on land and water. With all this going for it, the CSG industry is the 800-pound gorilla – it goes pretty much where and does pretty much what it wants. There is no national land-use policy that might give some geographical direction to prioritising deposits to exploit, and dominance of subsurface rights means opportunity costs on the surface do not weigh heavily in industry calculations. The CSG industry has, in fact, focused on some of Australia's relatively scarce prime farmland, especially in the Surat and Bowen basins of Queensland.

At this stage in its history, Australia is not wondering whether to let CSG in the door. The facts on the ground are that CSG is so dominant economically that it has imposed, pretty much on its own terms, co-existence upon agriculture, even on prime farmland. As a practical matter, the economic question now must be posed in terms of "what, if any, countervailing considerations might challenge the facts on the ground?". The obvious countervailing considerations are the environmental externalities, the long-run impacts of an exhaustible-resource-extraction enterprise with a clear expiration date, and the leakage (from an Australian perspective) of resource rents due to dominant foreign ownership of the industry. The economic question is whether, and under what conditions, these countervailing factors might change the bottom line.

In what follows I summarise the reasoning and some of the empirical findings of Chen and Randall,[1] who addressed this question in a case study of Arrow Energy's Surat basin CSG project. The project covers an area of approximately 8600 square kilometers (3320 square miles) in the Darling Downs, a region renowned for its agricultural productivity. Sixty percent of the project area is considered prime cropland and most of the remainder is suitable for livestock grazing.[68] While Arrow Energy is perhaps an unfamiliar entity, it is a partnership of two very prominent organisations: Shell Oil and the government of mainland China.

A benefit cost analysis (BCA) framework was used to assess the absolute and relative economic net benefits of CSG and agriculture. This involved computing and monetising the relevant benefits and costs to calculate the net present value (NPV), per hectare and for the whole project area, generated by CSG mining alone, agriculture alone, or both co-existing on the same piece of land. The time-paths of annual net benefits from CSG and agriculture differ quite sharply (see Figure 5). Agriculture, being already well established, delivers positive net benefits from the outset and benefits are expected to increase gradually in real terms over time.[69] CSG has high up-front costs, a period of high and fairly steady net benefits lasting around 30 years, and a stream of continuing environmental costs after the gas is gone. The level of these continuing costs depends on the completeness and effectiveness of decommissioning the wells and removing the infrastructure. Under these circumstances, the 30- to 50-year time horizons customarily used in BCA would capture the full economic benefits of CSG development while ignoring the continuing environmental damage after the gas is exhausted and thus distort the results in favour of CSG. The process of re-equilibrating the interconnected ground- and surface-water systems could well take several hundred years. Because little confidence could be placed in estimates of key variables so far into the future, NPVs were calculated over a 100-year time horizon.

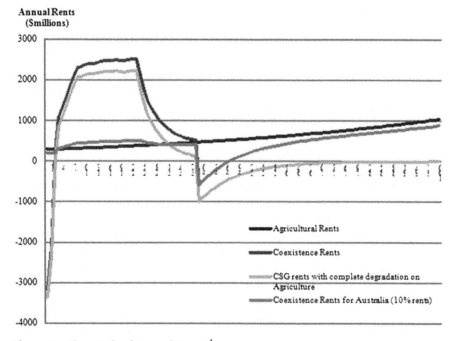

**Figure 5**   Time Path of Annual Rents.[1]

As emphasised early in this chapter, many of the environmental impacts of CSG development are uncertain or unknown, and there is always uncertainty about the key economic variables, *e.g.* demands, production conditions and prices. For key variables whose values are unknown or known but subject to uncertainty, sensitivity analysis was conducted to identify the effects of a plausible range of values for these variables on the net benefits. Baseline (roughly, mid-range) estimates of key variables, the range of values used in sensitivity analysis, and data sources are documented.[1]

From a national perspective it is reasonable to evaluate CSG in terms not of its net value to the mostly international operators, but of the net value retained in Australia. While the royalties collected by state and territorial governments (nominally 10% *ad valorem*, but actual collections are somewhat less) are readily observable, the proportions of corporate tax receipts collected and earnings accruing to Australian stockholders in multi-national CSG operators that can be attributed to exhaustible resource rents are much more opaque.

Sensitivity analyses were conducted for six key variables: the price of CSG, the proportion of CSG rents captured in Australia, the rate of growth in value of agricultural output, the external environmental costs of CSG, the level of agricultural degradation caused by CSG extraction, and the discount rate applied to calculate NPV. In addition, three distinct cases were evaluated: agriculture alone, CSG alone, and co-existence of agriculture and CSG. To simplify exposition while delivering the key messages, we defined seven scenarios and calculated NPVs for four cases for each scenario (see Figure 6). The scenarios included two agriculture-only *vs.* CSG-only scenarios, with parameter values favourable for agriculture and baseline values, respectively; and five co-existence *vs.* agriculture-only scenarios, with parameter values favourable for agriculture, all values low, baseline values, all values high, and values favourable for CSG, respectively. The four cases were: CSG where all rents count, agriculture only, CSG where Australia retains 10% of rents, and CSG where Australia retains 30% of rents.

## 4.1 Results

The key results of this analysis include the following:

- In the case of agriculture and CSG co-existence, the key immediate impact of CSG on rents from agriculture is diminished agricultural productivity arising from loss of land and restricted mobility of farm machinery, both attributable to CSG wells and infrastructure.
- After the CSG has been depleted, the co-existence annual net benefits will always stay below the agriculture line as diminished agricultural production continues long into the future. This indicates the possibility that the net annual benefits gap between CSG mining and

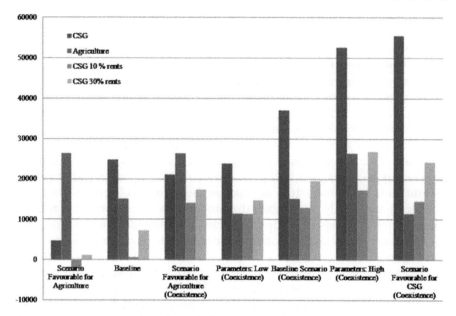

**Figure 6**   Net present values (NPVs) of seven scenarios.
(Source: C. Chen and A. Randall, The economic contest between coal seam
gas mining and agriculture on prime farmland: it may be closer than we
thought, *J. Econ. Soc. Pol.*, 2013, **15**(3), Art. 5).[1]

agriculture-only (Figure 5) may be closed, given enough time and a low-
enough rate of discount.

- CSG, either alone or in co-existence, generates greater NPV than agriculture in the baseline (*i.e.* mid-range) parameter scenarios, so long as
  all CSG rents count in the calculation.
- When only 10% of CSG rents count from an Australian perspective, the
  only scenario in which CSG generates more NPV than agriculture is that
  with parameter values systematically favouring CSG. If Australia retained 30% of the CSG rents and parameters were set at baseline values,
  CSG would have greater NPV in the co-existence case but not the CSG-
  alone case.
- When parameter values systematically favour agriculture, agriculture
  generates greater NPV than CSG (alone or in co-existence) even when all
  CSG rents count in the calculation.
- Given the uncertainties and unknown unknowns, Chen and Randall
  conceded the possibility that their worst-case analysis may have
  understated the environmental damage from CSG development.[1]

## 5   Conclusions

Given the advantages the CSG industry in Australia enjoys – buoyant export
demand for LNG and an accommodating minerals rights, taxation and

regulatory regime – it resembles the proverbial 800-pound gorilla. It goes pretty much where and does pretty much what it wants. To question whether it serves the Australian interest as well as it might is to ask whether there are serious countervailing considerations that could change to bottom line. These might include the environmental externalities, the long-run impacts of an exhaustible-resource-extraction enterprise with a clear expiration date, and the leakage (from an Australian perspective) of resource rents due to dominant foreign ownership of the industry.[1,70]

The Chen–Randall findings suggest several take-away messages. First, the overwhelming dominance of CSG in the Australian rural marketplace is not so robust when the full environmental costs of CSG development are approximated and incorporated into the analysis. In other words, markets, which seem to be offering unambiguous endorsement of CSG development in Australia, provide a seriously incomplete guide to CSG benefits and costs, especially those CSG benefits and costs that accrue to Australia.[1]

Second, of all the key parameters in this analysis, NPV is most sensitive to the price of LNG, a sobering thought given the potential entry of North American exporters into global export markets and the longer-range possibility that vast shale gas resources will developed.

Third, Australia's stake in its CSG industry depends crucially on its willingness to assert stronger claims to the resource rents generated. In this respect, recent political developments are not promising: the current government was elected in September 2013 on a platform that featured abolition of both the fledgling carbon price and a very tentative minerals resource rent tax.

Finally, one should not make too much of the Chen–Randall case study. It gets so much attention in this chapter because it is unique, to my knowledge, in presenting a numerical, empirically based benefit–cost evaluation of CSG development on prime Australian farmland. Nevertheless, it is not definitive and the authors were transparent about the data limitations and the resort to sensitivity analysis when faced with key variables of unknown and/or uncertain magnitudes.

Uncertainties and unknowns are in fact the big story. On-going and planned CSG development is of such magnitude that its cumulative impacts on such a vast and complex system as Australia's Great Artesian basin are unknown and perhaps unknowable in advance.

## References

1. C. Chen and A. Randall, The economic contest between coal seam gas mining and agriculture on prime farmland: it may be closer than we thought, *J. Econ. Soc. Pol.*, 2013, **15**(3), Art. 5.
2. British Petroleum, *Energy Outlook*, 2013; http://www.bp.com/en/global/corporate/press/press-releases/bp-energy-outlook-2030-shows-increasing-impact-of-unconventional-oil-and-gas-on-global-energy-markets.html, (last accessed 24/06/2014).

3. International Energy Agency, *World Energy Outlook, 2013*; http://www.iea.org/media/files/WEO2013_factsheets.pdf (last accessed 24/06/2014).
4. Tony Abbott's move to axe the carbon tax has been applauded by Canada's Harper government, which rejected a similar policy in 2008, *The Australian*, 13 November 2013; http://www.theaustralian.com.au/national-affairs/in-depth/carbon-tax-repeal-bill-is-prime-minister-tony-abbotts-first-item/story-fndttws1-1226758826441 (last accessed 24/06/2014).
5. Bureau of Resources and Energy Economics, *Energy and Resources Statistics*, 2013, Canberra.
6. D. Kahan, P. Slovic, D. Braman and J. Gastil, Fear of democracy: a cultural evaluation of Sunstein on risk, *Harvard Law Rev.*, 2006, **119**, 1071.
7. W. M. Corden, Dutch disease in Australia: policy options for a three-speed economy, *Austral. Econ. Rev.*, 2012, **45**(3), 290.
8. R. Gregory, "Living standards, terms of trade and foreign ownership: reflections on the Australian mining boom", *Aust. J. Agri. Resour. Econ.*, 2012, **56**(2), 171.
9. R. Solow, Intergenerational equity and exhaustible resources, *Rev. Econ. Stud.*, 1974, **41**, 29–45.
10. J. Hartwick, Intergenerational equity and the investing of rents from exhaustible resources, *Am. Econ. Rev.*, 1977, **67**, 972.
11. Reserve Bank of Australia, *Statement on Monetary Policy. Box B: The Mining Sector and the External Accounts*, 2011; www.rba.gov.au/publications/smp/boxes/2011/nov/b.pdf (last accessed 24/06/2014).
12. L. Hogan and R. McCallum, *Non-renewable Resource Taxation in Australia*, Report prepared for AFTS Review Panel, Australian Bureau of Agricultural and Resource Economics, Canberra, October 2010.
13. D. Rosenau-Tornow, P. Buchholz, A. Riemann and M. Wagner, Assessing the long-term supply risks for mineral raw materials – a combined evaluation of past and future trends, *Resour. Pol.*, 2009, **34**(4), 161–175.
14. D. Jacks, *From Boom to Bust: A Typology of Real Commodity Prices in the Long Run*, NPER Working Paper 18874, 2013, National Bureau of Economic Research, Cambridge MA.
15. Bureau of Resources and Energy Economics, *Gas Market Report*, 2013, Canberra.
16. W. Schulze, The optimal use of non-renewable resources: The theory of extraction, *J. Environ. Econ. Manage.*, 1974, **1**(1), 53–74.
17. P. Dasgupta, G. Heal and J. Stiglitz, *The Taxation of Exhaustible Resources*, Working paper 436, National Bureau of Economic Research, Cambridge MA, 1980.
18. M. Hunt, *Mining Law in Western Australia*, 4th ed, Leichhardt, NSW, The Federation Press.
19. T. Bergstrom, Property rights and taxation in the Australian minerals sector, *Selected Works of Ted C. Bergstrom*, University of California Santa Barbara, 1984; http://works.bepress.com/ted_bergstrom/93 (last accessed 29/09/2013).

20. T. Daintith, *Finders Keepers? How the Law of Capture Shaped the World Oil Industry*, Washington, DC, RFF/Earthscan, 2010.
21. Ashurst, *Mining in Australia - an introduction for investors*, http://www.ashurst.com/doc.aspx?id_Content=8821 (last accessed 24/06/2014).
22. T. Taggart, *The Free Entry Mineral Allocation System in Canada's North: Economics and Alternatives*, Working Paper 6, Canadian Arctic Resources Commission, 1998.
23. Standing Council on Energy and Resources, *The National Harmonised Regulatory Framework for Natural Gas from Coal Seams 2013*, Canberra, May 2013; http://scer.govspace.gov.au/files/2013/06/National-Harmonised-Regulatory-Framework-for-Natural-Gas-from-Coal-Seams.pdf (last accessed 24/06/2014).
24. W. Carlisle, Australian Broadcasting Corporation, *Coal Seam Gas: By the Numbers*, 2012; http://www.abc.net.au/news/specials/coal-seam-gas-by-the-numbers/ (last accessed 24/06/2014).
25. ACIL Tasman, *Economic Significance of Coal Seam Gas in Queensland*, Response to the call for input from APPEA, 2012, Australian Council of Learned Academies, Securing Australia's Future: Project Six Engineering Energy: Unconventional Gas Production, 2012; . http://www.appea.com.au/wp-content/uploads/2013/05/120606_ACIL-qld-csg-final-report.pdf (last accessed 24/06/2014).
26. P. Cook, V. Beck, D. Brereton, R. Clark, B. Fisher, S. Kentish, J. Toomey and J. Williams, *Engineering Energy: Unconventional Gas Production*, Report for the Australian Council of Learned Academies, 2013; www.acola.org.au (last accessed 24/06/2014).
27. J. Williams, A. Milligan and T. Stubbs, *Coal Seam Gas Production: Challenges and Opportunities*, Gas Market Report, Bureau of Resource and Energy Economics, Canberra, 2013, 42.
28. Queensland Department of Environment and Heritage, *Fraccing*, http://www.ehp.qld.gov.au/management/non-mining/fraccing.html (last accessed 24/06/2014).
29. University of Technology, Sydney, *Drilling Down: Coal Seam Gas, A Background Paper*, for the City of Sydney, 2011, Institute for Sustainable Futures.
30. G. Batley and R. Kookana, Environmental issues associated with coal seam gas recovery: managing the fracking boom, *Environ. Chem.*, 2012, **9**, 425–428.
31. D. Healy, *Hydraulic Fracturing or 'Fracking': A Short Summary of Current Knowledge and Potential Environmental Impacts*, University of Aberdeen, 2012; http://www.epa.ie/pubs/reports/research/sss/UniAberdeen_Fracking Report.pdf (last accessed 24/06/2014).
32. A. Randall, Coal seam gas – toward a risk management framework for a novel intervention, *Environ. Plan. Law J.*, 2012, **29**(2), 152.
33. National Water Commission, *Coal Seam Gas Update – June 2012*, National Water Commission, Canberra, 2012; http://nwc.gov.au/nwi/position-statements/coal-seam-gas (last accessed 24/06/2014).

*Alan Randall*

34. National Water Commission, *Onshore Co-Produced Water Extent and Management*, Waterlines Report 54, 2011.
35. Department of Sustainability, Environment, Water, Population and Communities, *Water Group Advice on EPBC Act Referrals, (September) for Minister Tony Burke*, Tabled in the Australian Senate, Canberra, 2010.
36. A. Randall, *Risk and Precaution*, Cambridge University Press, London and New York, 2011.
37. Queensland Water Commission, *Underground Water Impact Report for the Surat Cumulative Management Area*, Office of Groundwater Impact Assessment, Draft Report, Queensland Government Department of Natural Resources and Mines, Brisbane, May 2012; http://dnrm.qld.gov.au/__data/assets/pdf_file/0016/31327/underground-water-impact-report.pdf (last accessed 24/06/2014).
38. N. Swayne, Regulating coal seam gas in Queensland: lessons in an adaptive environmental management approach?, *Environ. Plan. Law J.*, 2012, **29**(2), 163.
39. J. Petesic, *It all starts underground: Estimating the Additional Costs of Irrigating Imposed by Coal Seam Gas*, Honours Research Report, unpublished, University of Sydney, 2013.
40. The Royal Society and the Royal Academy of Engineering, *Shale Gas Extraction in the UK: A Review of Hydraulic Fracturing, London*, June 2012.
41. D. Gerard, The law and economics of reclamation bonds, *Resourc. Pol.*, 2000, **26**(4), 189.
42. S. Vink, N. Kunz, D. Barrett and C. Moran, *Groundwater Impacts of Coal Seam Gas Development – Assessment and Monitoring*, Scoping study prepared for Centre for Water in the Minerals Industry, Sustainable Minerals Institute, University of Queensland, Brisbane, Australia, P08-010-002.doc, 2008.
43. C. Moran and S. Vink, *Assessment of Impacts of the Proposed Coal Seam Gas Operations on Surface and Groundwater Systems in the Murray-Darling Basin*, Centre for Water in the Minerals Industry, Sustainable Minerals Institute, The University of Queensland, 2010.
44. GHD, *Desalination in Queensland*. QLD Department of Natural Resources and Mines, 2003.
45. Santos, *Coal Seam Gas and Water*; http://www.santos.com/coal-seam-gas/coal-seam-gas-water.aspx (last accessed 24/06/2014).
46. T. Clark, R. Hynes and P. Mariotti, *Greenhouse Gas Emissions Study of Australian CSG to LNG*, Worley-Parsons Resources and Energy, 2011; http://www.abc.net.au/radionational/linkableblob/4421188/data/greenhouse-gas-emissions-study-of-australian-csg-to-lng-data.pdf (last accessed 24/06/2014).
47. L. Cathles, M. Brown, M. Taam and A. Hunter, A Commentary on 'The greenhouse gas footprint of natural gas in shale formations' by R.W. Howarth, R. Santoro and A. Ingraffea, *Clim. Change*, 2012, **113**, 525.
48. Department of Climate Change and Energy Efficiency, *Coal Seam Gas: Estimation and Reporting of Greenhouse Gas Emissions*; http://www.

climatechange.gov.au/sites/climatechange/files/files/consultations/NGA-FactSheet-7-CoalSeamGas-20120430-PDF.pdf (last accessed 24/06/2014).

49. Department of the Environment, *Reporting of Fugitive Greenhouse Gas Emissions under the NGER Measurement Determination, Technical Discussion Paper*, Canberra, April 2013.

50. R. Alvareza, S. Pacalab, J. Winebrakec, W. Chameides and S. Hamburg, Greater focus needed on methane leakage from natural gas infrastructure, *Proc. Natl. Acad. Sci. U. S. A.*, 2012, **109**(17), 6437.

51. L. Ries, R. Fletcher Jr, J. Battin and T. Sisk, Ecological responses to habitat edges: mechanisms, models, and variability explained, *Ann. Rev. Ecol. Evol. Syst.*, 2004, **35**, 491.

52. S. Cushman, Effects of habitat loss and fragmentation on amphibians: a review and prospectus, *Biol. Conserv.*, 2006, **128**(2), 231.

53. J. Fischer and D. Lindenmayer, Landscape modification and habitat fragmentation: a synthesis, *Global Ecol. Biogeog.*, 2007, **16**, 265.

54. M. Hansen and A. Clevenger, The influence of disturbance and habitat on the presence of non-native plant species along transport corridors, *Biol. Conserv.*, 2005, **125**, 249.

55. *Lock the Gate Alliance*; http://www.lockthegate.org.au/groups (last accessed 24/06/2014).

56. P. Dart, Coal Seam Gas a Risk to Food Security, *The Conversation*, May 24, 2011; http://theconversation.com/coal-seam-gas-a-risk-to-food-security-485 (last accessed 24/06/2014).

57. C. McCoy, Australia coy in report on heritage status of Great Barrier Reef, *The Conversation*, 2013; http://theconversation.com/australia-coy-in-report-on-heritage-status-of-great-barrier-reef-11916 (last accessed 24/06/2014).

58. United Nations Education Scientific and Cultural Organization, *Committee Decisions, 36COM 7B.8 Great Barrier Reef (Australia) (N 154)*, 2012; http://whc.unesco.org/en/decisions/4657/ (last accessed 24/06/2014).

59. A. Ricketts, Investment risk: an amplification tool for social movement campaigns globally and locally, *J. Econ. Soc. Pol.*, 2013, **15**(3), Art. 4.

60. T. Poisel, Coal seam gas exploration and production in New South Wales: the case for better strategic planning and more stringent regulation, *Environ. Plan. Law J.*, 2012, **29**(2), 129.

61. D. Lloyd, H. Luke and W. Boyd, Community perspectives of natural resource extraction: coal-seam gas mining and social identity in Eastern Australia, 2013, *Coolabah* 10, 144; http://epubs.scu.edu.au/cgi/viewcontent.cgi?article = 2476&context = esm_pubs (last accessed 24/06/2014).

62. Corrs Chambers Westgarth, *Proposed Coal Seam Gas Exclusion Zones in NSW*, 2013; http://www.corrs.com.au/publications/corrs-in-brief/proposed-coal-seam-gas-exclusion-zones-in-nsw/ (last accessed 24/06/2014).

63. F. Haslam-McKenzie, J. Rolfe, A. Hoath, A. Buckley and L. Greer, *Regions in Transition: Uneasy Transitions to a Diversified Economy involving Agriculture and Mining*, Report to CSIRO Minerals Down Under Flagship, Mineral Futures Collaboration Cluster, 2013.

64. A. Walton, R. McCrea, R. Leonard and R. Williams, Resilience in a changing community landscape of coal seam gas: Chinchilla in southern Queensland, *J. Econ. Soc. Pol.*, 2013, **15**(3), Art. 2.

65. Queensland Department State Development, Infrastructure and Planning, *Social Impact Assessment Guideline*. Coordinator-General, Brisbane; http://www.dsdip.qld.gov.au/resources/guideline/social-impact-assessment-guideline.pdf (last accessed 24/06/2014).

66. K. de Rijke, Coal seam gas and social impact assessment: an anthropological contribution to current debates and practices, *J. Econ. Soc. Pol.*, 2013, **15**(3), Art. 3.

67. Santos, Vital Roma worker accommodation nears completion, 2013; http://www.santosglng.com/resource-library/news-announcement/vital-roma-worker-accomodation-nears-completion.aspx (last accessed 24/06/2014).

68. Coffey Environments, *Surat Gas Project Environmental Impact Statement*, Arrow Energy, 2012, Brisbane.

69. Australian Bureau of Agricultural and Resource Economics and Sciences, *Food Demand to 2050 – Opportunities for Australian Agriculture*, 2012, Canberra.

70. T. Wood, L. Carter, and D. Mullerworth, *Getting Gas Right: Australia's Energy Challenge*, Grattan Institute, Melbourne, 2013.

# Prospects for Shale Gas Development in China

SHU JIANG

## ABSTRACT

China has the largest estimated quantity of recoverable shale gas in the world. Inspired by the shale gas revolution in the US, China is trying to replicate the successful US experience to produce shale gas to power its economy, which is the second largest in the world, and mitigate the pressure felt from being the world's biggest greenhouse gas emitter. Both optimistic and pessimistic prospects for China shale gas development have been debated in the last several years. The promising prospect of shale gas development in China is based on having the largest shale gas resources in the world, huge internal demand, the government's supportive policy and an established preliminary commercial production. The challenges for China shale gas include: the complex geology in marine, transitional and lacustrine settings; lack of advanced shale gas technologies and experience; insufficient infrastructure; environmental issues, especially the complex tectonic and stress conditions; high clay mineral content for lacustrine shale; harsh ground conditions; and water shortages. These issues have meant that US hydraulic fracturing experiences do not work well for China's shales. However, shale gas provides an indispensable opportunity for China to develop this enormous resources to secure its energy supply gap; power its fast-growing economy; create socioeconomic and employment benefits; mitigate pollution pressure; and increase geopolitical leverage. Increasing knowledge of China's shale geology; Sino-foreign joint exploration; shale gas technologies as commodity; indigenous

Issues in Environmental Science and Technology, 39
Fracking
Edited by R.E. Hester and R.M. Harrison
© The Royal Society of Chemistry 2015
Published by the Royal Society of Chemistry, www.rsc.org

technologies; and the recent breakthrough of commercial shale gas production in Sichuan Basin have made China's future shale gas plan possible. The full financial and policy support from government to import US equipment and technologies, improve infrastructure, liberalise land and markets for small private companies with capital will facilitate the take-off of shale gas development in China.

# 1 Introduction

The surge in shale gas production which has resulted from technological advances in hydraulic fracturing, horizontal drilling, geological characterisation and shale property measurement has revolutionised the US energy scene.[1] This shale gas bonanza helped the US to surpass Russia to become the world's largest natural gas producer in 2009. It may result in the US displacing Saudi Arabia as the world's largest oil producer around 2020 and securing energy self-sufficiency.[2] Shale gas has also resulted in the US shifting from a net natural gas importer to a potential net natural gas exporter, has driven US gas prices down and has benefited the US economy by creating new jobs and helping manufacturing and related industries. The dramatic switch from coal to cleaner natural gas provides a paradigm shift to reduce greenhouse gas emissions. The rest of the world, especially China, may possibly replicate this success to reshape the global energy market.

China is the world's most populous country, its largest energy consumer, its second largest economy and the biggest emitter of greenhouse gases. In the last decade, China has been relying on importing energy to fuel its fast-growing economy. China has experienced energy consumption growth of 136% between 2001 and 2011.[3] At present, coal makes up 70% of the fuel needed to power China's economy. As a consequence of coal use, air pollution causes the Chinese government serious headache because of the outcry caused by smog pollution. China has developed ambitious plans for alternative renewable energy, but at the current time renewables are still very expensive and these comprise only a small portion of total consumption. Nuclear energy also is perceived to carry a lot of risks due to the recent incident of contamination resulting from Japan's Fukushima reactor accident.

Natural gas provides the cleanest hydrocarbons for power generation, transportation, chemical feedstocks, and industrial and domestic fuel. The displacement of coal by gas in power generation results in the release of up to 50% less carbon dioxide emissions.[3] However, China's natural gas consumption only accounts for 4% of its energy mix compared to the global average of 24%, due to its huge energy demand and limited natural gas supply.[4] In order to sustain economic development and to secure an abundant energy supply and lower $CO_2$ emissions, cleaner natural gas has been chosen as the best alternative to inefficient coal in China. Right now, domestic gas production cannot satisfy industrial and residential needs. By 2020, China's natural gas consumption will reach 380 billion cubic metres and the demand gap will be 180 billion cubic metres, since the conventional gas production will be only 200 billion cubic metres at that time (Dawei Zhang, Ministry of Land and Resources of China, personal communication, June 14, 2013). Fortunately, China is sitting on the largest shale gas reserves in the whole world. The exploration and development of these reserves have potentially wide ranges of benefits for China's economy, energy security and mitigation of $CO_2$ emission. China is looking to imitate a decade-long growth of the commercial shale industry in the United States, which grew from 2 to 23% of total gas production from 2001 to 2010.[5] So far, exploration of shale gas is still at an early stage in China and there have been many doubts expressed over the government's seemingly unrealistic 12th Five-year Plan target of 6.5 billion cubic metres (230 billion cubic feet) per year of shale gas production by 2015, due to the slow progress over the past several years.[4] One recent breakthrough which may change the skeptics' attitude and make China's 2015 output target possible is based on the commercially viable production of shale gas in Sichuan Basin, especially in the Fuling area recently in late 2013, which makes China the only country outside of North America that has reported commercial production.

At the same time as reporting shale gas exploration progress, Beijing adopted new policies and announced financial incentives for the development of the shale gas industry, as well as measures to encourage competition and liberalization of the natural gas market. Generally, both encouraging prospects and challenges exist for China's shale gas development. The purpose of this chapter is to review the shale gas resources in China and to examine the pros and cons for shale gas development in China.

## 2 Geology and Shale Gas Resources in China

The organic rich shales spanning from the Pre-Cambrian Sinian period (prior to the Cambrian period and 700 million years before the present day) to the Quaternary period (the last two million years up to the present day) are widely distributed in China. The Pre-Cambrian to Upper Paleozoic organic-rich shales with gas window maturity; Mesozoic to Cenozoic organic-rich shales with wet gas window maturity; and shallow Quaternary shales with biogenic gas all have shale gas potential. The Pre-Cambrian to Lower Paleozoic shales are distributed all over China and were deposited in passive margin to foreland marine settings; the Upper Paleozoic (Carboniferous to Permian) shales, mainly in South China, North China and NW China, were deposited in transitional (coastal swamp associated with coal) cratonic settings; the Meso-Cenozoic sporadically distributed shales were deposited in rift, lacustrine (lake) settings (see Figure 1). Table 1 illustrates the wide distribution of major potential gas shales in time and space for all the China shales in three depositional settings. Comparing China gas shale resource basins with US gas shale resource basins, the primary difference is that the potential shale gas basins in China are widely distributed in various tectonic and sedimentary settings throughout the Chinese territory, whereas US shale

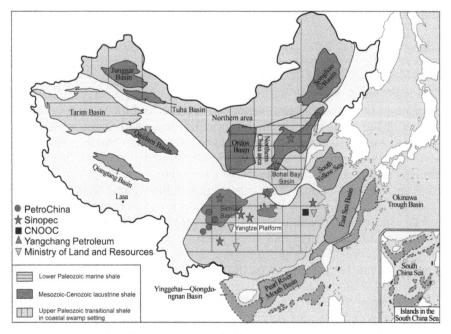

**Figure 1**   The exploration activities for shales of various ages in different depositional settings in China.
(Base map modified from Zou *et al.* in 2011).[6]

basins are mainly distributed in the Appalachian thrust belt in the east and
the Rocky Mountain thrust belt in the west and are defined by marine depositional environments.

In 2010, the Strategic Research Center of Oil and Gas – Ministry of Land and
Resources (MLR) and China University of Geosciences at Beijing used an
analog assessment method to announce that China Shale Gas resources are
predicted to be about 30 trillion cubic metres (TCM) or 1059 trillion cubic feet
(TCF). In 2011, the US Energy Information Administration (EIA) assessed that
China could have 36.11 TCM or 1275 TCF of technically recoverable shale gas.
In March 2012, the China Ministry of Land and Resources announced that
China had 25.08 TCM (886 TCF) of recoverable shale gas reserve in marine,
transitional and lacustrine geologic settings in onshore China (excluding the
Tibetan Plateau area). Recently, in June 2013, EIA reduced their assessment of
China's recoverable shale gas reserve a little to 31.56 TCM or 1115 TCF.
Either number indicates that China's shale resources are the largest in the
world and comparable with that of the US's updated 665 TCF recoverable
shale gas. In terms of shale gas proportions in different geologic ages, the
Paleozoic period dominates with 66.7%, followed by the Mesozoic 26.7%
and Cenozoic 6.6%. For the different depositional settings, the marine and
transitional account for 68.41% and lacustrine shale gas has 31.59%.[7] These
figures indicate that both marine and lacustrine shales in various

**Table 1** China gas shale depositional settings and their distribution in time and space.

| Depositional setting | Age and formation | | Distribution area |
|---|---|---|---|
| Lacustrine | Cenozoic | Neogene | Qaidam Basin |
| | | Paleogene | Bohai Bay Basin, Qaidam Basin |
| | Mesozoic | Cretaceous | Songliao Basin |
| | | Jurassic | Turpan-Hami, Junggar, Tarim, Qaidam, Sichuan Basin |
| | | Triassic | Ordos Basin, Sichuan Basin |
| | Paleozoic | Late Permian | Junggar, Turpan-Hami |
| Transitional (coastal setting associated with coal) | Paleozoic | Late Permian (Longtan Fm) | Yangtze Platform including Sichuan in Upper Yangtze |
| | | Early Permian (Taiyuan, Shanxi Fm) | North China |
| | | Late Carboniferous (Benxi Fm) | North China |
| Marine | Paleozoic | Early Silurian (Longmaxi Fm) | Yangtze Platform including Sichuan in Upper Yangtze |
| | | Late Ordovician (Wufeng Fm) | Yangtze Platform including Sichuan in Upper Yangtze, Tarim Basin |
| | | Early Cambrian (*e.g.* Qiongzhusi Fm) | Yangtze Platform, Tarim Basin |
| | Pre-Cambrian | Sinian (*e.g.* Doushantuo Fm) | Upper and Middle Yangtze Platform |

stratigraphic intervals are important in China. The unconventional shale gas reserves in China are ten times greater than the current proven conventional gas reserves. These huge shale gas resources represent a vast, long-term and very important energy source for China, which suggests there is a great opportunity to emulate the shale gas revolution in the U.S.

Geological investigations for resource assessment and exploration show that organic-rich marine shales in China have good shale gas play potentials based on high organic matter content, high maturity, high content of brittle minerals (see Figure 2) and high intra-organic nano-porosity (see Figure 3). For example, the Qiongzhusi shale of the Lower Cambrian in Yangtze Platform, including the Sichuan Basin, has an average thickness of 120 metres over the region with average total organic carbon (TOC) of 2.8% and average vitrinite reflectance (Ro, a measure of the thermal history of hydrocarbon source sediments) of 3%. The Longmaxi formation shale of the Lower Silurian has an average thickness of 100 metres across the region with an average TOC of 2.6% and Ro between 2% and 3%. Both of these two typical

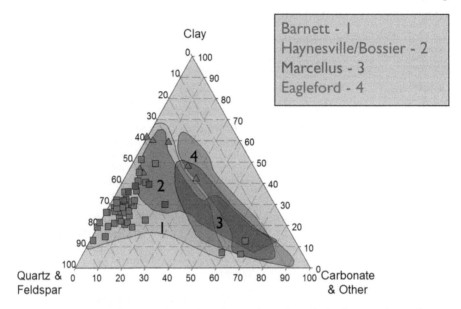

**Figure 2**    Ternary diagram for mineralogy of marine shale (square legend) and
lacustrine shale (triangle legend) in China and its comparison with
mineralogy of typical US shales (number legend).

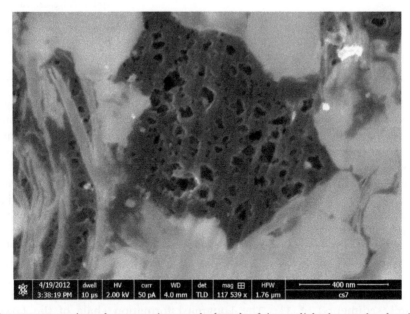

**Figure 3**    Scanning electron micrograph (SEM) of ion-polished sample showing
abundant intra-organic nano-pores in Silurian marine shale from the
Sichuan Basin, China.

marine shales have brittle quartz and feldspar content greater than 40% (see Figure 2). Generally, China lacustrine shales have higher clay content than marine shales (see Figure 2). This is why many experts currently think it is much more difficult to frac the lacustrine shale.

## 3 Recent Progress of Shale Gas Development in China

China has performed shale gas resource assessments and conducted exploration for nearly six years. In order to encourage shale gas exploration, the Ministry of Land and Resources (MLR) has held two separate shale gas tenders – four shale blocks in 2011 and 19 blocks in 2012 – which attracted nearly 20 companies to join in shale gas exploration. The junior companies include central and provincial level state-owned enterprises and private companies. So far, 300 shale gas wells (including shallow parameter wells behind outcrops) targeting marine, lacustrine and transitional shales have been drilled by PetroChina, Sinopec, CNOOC, Yanchang Petroleum, other state or private companies recently allocated shale blocks, MLR and foreign partners of Chinese companies. The exploration activities have been mainly focused in the Sichuan Basin for marine and lacustrine shale gas; the Yangtze Platform outside the Sichuan Basin for marine shale gas; the Ordos Basin for lacustrine shale gas; the Bohai Bay Basin and the Nanxiang Basin for lacustrine shale gas and shale oil (see Figure 1). Almost half of the drilled wells have shown good shale gas and/or shale oil. Several wells targeting marine shales were reported to be very successful based on test results, *e.g.* the Yang201-H2 well drilled by PetroChina and Shell in the Sichuan Basin was reported to produce at 430 000 cubic metres per day and recently the Jiaoye8-2HF well in the Fuling area in East Chongqiong was reported to commercially produce 547 000 cubic metres (19.1 million cubic feet) per day. Sinope drilled more than 30 shale gas wells in the Fuling pilot area and 6 wells targeting marine shale were reported to have a high rate commercial production with an average of 180 000 cubic metres per well. This breakthrough in the Fuling Pilot area in Chongqing makes Sinopec's 2015 output target in this area doubled. For lacustrine shale, Sinopec and Yanchang Petroleum recently sped up lacustrine shale gas exploration in the Sichuan Basin and the Ordos Basin, *e.g.* Sinopec tapped shale gas from the Jurassic lacustrine shale intervals in wells located in the Jiannan area in the eastern Sichuan Basin and the Yuanbe area in Northeastern Sichuan Basin; the Jian 111 well in the Jiannan area produced more than 500 000 $m^3$ in 2011 at a rate of 2000 $m^3$ per day. These recent significant strides may make China's 2015 output target of 6.5 BCM feasible and the target shale gas amount may even be doubled.

## 4 Socioeconomic and Employment Benefits of Shale Gas Development in China

In the US, the economic effects of the shale gas revolution have significantly changed the energy structure of the country and triggered a new round of socioeconomic development – lowering the gas price for households and industry, creating hundreds of thousands of jobs and adding billions to state and federal coffers for GDP growth.[8] The US natural gas supply has increased by 28% since 2005, and in 2011 about a third of domestic production came from shale gas. In 2010, prices at Henry Hub4 were less than $5.00 per million British thermal units (MBTU) and it was $2.46 in February 2012. This decreasing price trend will result in more than $2000 additional disposable income per household in 2035.[8] These price gaps are also giving energy-intensive US manufacturing industries, like petrochemicals, a competitive edge and making the US an attractive location for capital investment, which will pull US manufacturing industry out of recession and give a strong boost to the economic recovery. Shale gas supported more than 600 000 direct, indirect and induced jobs (*i.e.* jobs generated by the re-spending of received income resulting from direct and indirect job creation in the affected region). As the share of shale gas production increases to 60% in 2035, development of this resource will support more than 1.6 million jobs.[8] The oversupply of natural gas also makes the US energy self-sufficient and will create downward pressure on gas prices across the globe if the US exports its surplus shale gas to other markets, like the EU or Japan where the market price is much higher.

For China, copying the shale gas development from the US will reduce the need for gas imports and satisfy the gas demand needed to continue powering the booming economy. China's natural gas price is currently four times more than that of the US. Shale gas development will be good for companies in China; it will help the oil industry and related industries in terms of manufacturing, oilfield service and gas-fired power generation, create more employment and reduce gas prices for households and commercial clients. This appears to provide a cheaper route to a lower carbon economy than high-cost renewables. For the oil industry, the Chinese oilfield services companies, including Yantai Jereh Oilfield Services Group Co, SPT Energy Group, Kingdream Public Limited Company, Anton Oilfield Services, Honghua Group and others, have started manufacturing shale-gas drilling and fracturing equipment and other services for shale gas exploration. Recently, the Chinese Jereh's equipment for shale gas has been tested successfully in difficult topographic conditions and the company's shale gas fracturing equipment has been exported to the US since 2011. China's shale gas exploration also provides many opportunities for US service companies to supply tools, technology services, *etc.* Many of these companies' shale gas concept shares continue to surge. International service companies, like Halliburton, Schlumberger, Weatherford, *etc.*, have sold many shale gas hydraulic fracturing tools and provided fracturing services for horizontal shale gas wells in many of PetroChina and Sinopec's recent shale gas wells. There will also be other opportunities in related industries and sectors, such as downstream uses, *e.g.* gas-fired power generation, feedstock fuel. A better economy powered by clean shale gas and new business opportunities in many of the above-mentioned industries related to shale gas will help populous China to increase its employment rate. PetroChina, Sinopec, CNOOC and many private energy companies have set up shale gas research centers, subsidiary companies and manufacturing divisions which employ many technical people. If the shale gas development in China is as successful as in the US, it will help China's GDP to keep growing at a remarkable rate and build a resource-conserving and environmentally-friendly society to fulfil the Chinese government's commitment that energy consumption per unit of GDP should decrease by 16% while $CO_2$ emissions per unit of GDP should fall by 17% during the period from 2011 to 2015 according to China's 12th Five-Year Plan.

## 5  Geopolitical Implications of Shale Gas Development in China

The US-initiated shale gas revolution will not only change the global landscape of energy distribution but will also change the world's geopolitical layout. As unrest in the Middle East continues and US shale gas technologies become commercialised, the global landscape of energy resources will change, *e.g.* reducing requirements for imported liquefied natural gas (LNG). The US will become less and less reliant on Middle Eastern oil and take a more dominant position in the global energy distribution. The development of shale gas technology will have a major impact on the world's geopolitical map.

Geopolitics is affected by the trajectory of shale gas markets. The US's geopolitical pivot towards the Pacific region dovetails with shale revolution dynamics, which will help strengthen cooperation between the US and energy-consuming Asian nations, especially China. Indeed, the Russians have become anxious about the potential for exports to Asia from the US, Australia and East Africa. A geopolitical shift can be foreseen if a Pacific Rim energy supply chain stretching from Alaska, Canada, mainland USA and Australia to Japan overtakes the southern oil shipping route from the Middle East to Japan through the Malacca Strait and South China Sea, and this will be of strategic significance.[9] Fostering cooperation while exploring shale gas is of vital importance to both the US and China, which will help bring about a balanced development of the world economy and global geopolitical stability.

Since 2010, China's national oil companies (NOCs) have spent a lot of money to acquire upstream assets in the US, most of which has been spent on unconventional projects. China's NOCs have spent more money to acquire assets in North America ($44.2 billion) than in any other region of the world.[11] At the same time, as US oil companies sell assets abroad to earn money to invest in shale energy development in the United States, there are more assets available for China's NOCs to buy in other countries. One of the major geopolitical benefits of the US shale energy revolution is that it provides China's NOCs with a powerful incentive to curb their business with Iran, since the Committee on Foreign Investment in the United States has more leverage over the activities of Chinese companies who have invested in the US.

If the Chinese can bring their shale gas resources into line with shale gas in the US, it will give China more leverage in their natural gas pipeline negotiations with Russia. Therefore, shale gas could be a complete game changer with huge geopolitical implications. Transforming traditional importers of natural resources such as the US or China into net exporters can be a major trigger for massive change in international relations. When Russia and Persian Gulf countries lose their market share, a vital source of income and geopolitical advantage, they will have to find another source of economic growth and political power to prevent domestic uprising.

## 6  China's Supportive Policies for Shale Gas Development

The shale gas revolution in the US resulted from the confluence of factors, including: vast recoverable shale gas resources; access to shale gas resources on private lands; tax incentives from the government; economically attractive natural gas prices from 2007 to 2008; innovative technologies of horizontal drilling and hydraulic fracturing; and a good pipeline network, *etc.* The shale gas industry in China is still in early development, but the Chinese government has begun to provide strong political support to the shale gas industry in terms of their policies as a matter of national energy security. In 2009, Chinese president, Hu Jintao, and US president, Barak Obama, announced a US-China shale gas initiative at top government level to support China's shale gas development through learning from US experience.

Since broader social investment was previously precluded from natural gas extraction, shale gas blocks largely overlap with areas where PetroChina and Sinopec hold conventional oil and gas rights and, as shale gas development is not their priority, this makes it very difficult for other companies to enter the race. To remove historic barriers for shale gas development in China, many possible policies learned from the US and adaptations of China's national conditions have been discussed and reported.[10] In December 2011, shale gas was recognised as the 172nd mineral in China. This means that it can now be explored and extracted under an independent mineral right. It provides private capital with a chance to engage in shale gas. In 2012, the Ministry of Land and Resources offered a 2nd shale gas bidding and allowed even small private companies with capital to participate, which thus attracted more companies than in the 1st license round. At the same time, the Chinese energy regulator, the National Energy Administration (NEA), issued the *Shale Gas Industry Policy* in late 2013. The policy recommends certain reforms to encourage more companies to develop shale gas in China and promises to increase financial support for rapid shale gas exploration and development in a healthy and orderly manner. Based on this new policy, China's ambitious goal is to produce 6.5 billion cubic metres (BCM) annually (or 230 billion cubic feet) by 2015, and at least ten times that by 2020. This will be a huge leap from the estimated 2013 output of 200 million cubic metres. The following are the details of the supportive policies:

(1) To encourage Sino-foreign joint ventures in shale gas exploration and development in China. Foreign entities with advanced shale gas-related technologies are encouraged to partner with Chinese enterprises. In co-exploration, foreign companies provide advanced technologies and equipment for a fee. Several foreign oil and gas enterprises, *e.g.* Shell, Chevron, ConocoPhillips, ExxonMobil, Total, *etc.*, have already signed joint research agreements with Chinese partners. In March 2013, a shale gas production sharing agreement (PSA) between Shell and China National Petroleum Corporation (CNPC) was approved by the Chinese government, which is the first Sino-foreign PSA approved in China's shale gas industry.

(2) To encourage private and small companies with capital and flexible plans, since the shale gas boom and technology innovation in the US was led by small and medium-sized independent companies.

(3) To provide new financial incentives and government support.

According to the recent policy, tax incentives for the shale gas industry will borrow from the coal-bed methane (CBM) policy. The government will subsidise 0.4 CNY (Chinese currency: renminbi) per cubic metre for shale gas production from 2012 to 2015. This is twice as much as the current subsidy for CBM production. In addition, the local government may provide shale gas development and utilisation with appropriate subsidies according to the actual situation as it progresses. Also, the shale gas enterprises will get a waiver or reduction in license fees, mineral resources compensation fees, an exemption from value added taxes and an exemption from customs tariffs for equipment imported for shale gas exploration and development projects. For example, the government subsidised the Petro-China-Southwest subsidiary with 4.08 million CNY for 10.2 million cubic metres shale gas produced from four wells in the Weiyuan and Changning areas in the Sichuan Basin. At the same time, the National Development and Reform Commission (NDRC) and the National Energy Administration (NEA) have approved several shale gas pilot development areas (mostly within or around the Sichuan Basin). In these pilot areas, the Chinese government will facilitate the construction of supporting facilities and encourage joint collaboration of shale gas participants, technology integration, *etc.*

(4) To improve land and market access.

When reviewing an application for a land-use permit, a shale-gas developer will be given priority based on the National Development and Reform Commission's policy 2012.[12] Since China's existing gas pipeline network is monopolised mainly by CNPC and partially by Sinopec and is thus inaccessible to other companies, this has been a barrier to the development of shale gas. Accordingly, the new shale gas policy encourages private investment in the construction of new gas pipelines and infrastructure. Government also issued the mandatory requirement of "non-discriminatory infrastructure access" to pipelines. CNPC said it had started building the country's first shale gas pipeline at Changning in the Sichuan Basin and the pipeline in China will be extended to 150 000 km by 2020 (four times longer than its current length). This will help the delivery of shale gas in the future.

(5) To ensure environmental protection.

At the current time, an environmental assessment must be done when a shale gas well is planned and the drilling, hydraulic fracturing, waste disposal and emissions must meet environmental requirements. During hydraulic fracturing, the cementing process is monitored to ensure the integrity of the wellbore and prevent formation fluids from contaminating the shallow aquifer zone. After the treatment job, the drilling mud and fracturing fluids are required to be recycled.

# 7   Challenges for Shale Gas Development in China

China is looking forward to replicate the shale gas boom in the US and the outlook for the development and extraction of shale gas in China seems favorable. However, China still faces many formidable challenges related to geology, ground conditions, policies, technology, finance, water supply, infrastructure, *etc.* Generally speaking, there exist at least the following five major challenging categories:

## 7.1   Complex Geology

So far, both academia and industry lack fundamental understanding about China's shale geology, stress field (*i.e.* a region in a body for which the stress is defined at every point), and the technology suitable for Chinese shale fracturing and production.

The complex geologic and tectonic setting and different geomechanical regime in China's basins challenges many international companies with successful US shale experience attempting to fracture shales in China. Despite the rich reserves, the Chinese marine gas shale is either buried deeper than 3000 metres (over 5000 metres in West Sichuan Basin and the Tarim Basin) or faulted and folded or outcropped. The lacustrine shale is clay-rich and has rapid property change spatially. Historically, little shale geology data has been available and there has been little exploration activity. So basic geological elements are not yet well understood and it is difficult to locate the best plays. US shale-gas producing basins have relatively simpler geological and tectonic settings than those in China. The Chinese shales have more complex geologic settings than US shales in terms of depositional and tectonic settings and geological history. The promising marine shale in China is similar to the brittle Barnett shale in the US with regard to its mineralogy and gas content; the complex tectonic setting, much more complex diagenetic history and

harsh ground conditions caused by tectonic movements make shale gas extraction in China more challenging than that in the US. In particular, the
historical multi-stages of strong tectonic compression and extension in China
cause the shales to have different stress fields than those in the US, *e.g.* the
maximum principal stress is horizontal in some areas in the Sichuan Basin in
China while the maximum principal stress is vertical in US shale basins; this
is why the fracturing experiences in the US may not work very well in China.
Consequently, the drilling and stimulation cost in China is twice that of US
shales. For instance, the cost of a shale gas well drilled by CNPC in Sichuan is
$11 million in comparison to $4–6 million spent in the United States.[13] More
is needed to be learnt about the geology, geomechanics and hydraulic fracturing design for the unique China shales.

## 7.2 Lack of Advanced Shale Gas Development Technologies and Experience

So far, the technologies and equipment for developing shale gas is dominated by the US and China is still at an early stage in its shale gas development. Currently, China relies on cooperation with foreign enterprises with
US shale-producing experience. What's more, because of the complex geology of China shales, the technologies employed in US may not be a good fit
for the development of shale gas in China. In the US, the shale gas industry
was dominated by small innovative companies and the majority of work was
done by highly competitive service companies. While in China, the industry
was traditionally dominated by state-owned giant oil companies, they did
not have shale gas development technologies and shale gas was not their
priority. The Chinese companies who won blocks from the second licensing
round are mainly small companies or government investment bodies with
little or no direct experience in the oil and gas industry.

## 7.3 Infrastructure

An already existing and extensive pipeline network has enabled the US to
quickly bring its new shale gas bounty to market. China's pipeline network
was about 43 452 kilometres by the end of 2011, which is only about one
tenth of that of the US. There is also a lack of service infrastructure and a
supply chain to transport and manage equipment and the large quantities of
water needed to set up drilling operations in China. Since shale gas extraction requires immense amounts of water, an infrastructure has to be
created to make such quantities available.

## 7.4 Water Shortage in Some Areas

With about 20% of the world's population and only 6% of the world's water
resources, China is one of the least water-secure countries in the world.
China's renewable water resource *per capita* was estimated at 2063 m$^3$ per

year in 2011, only one fourth of the world average.[14] China's severe water shortages have already constrained economic growth during the last decade.[4] Hydraulic fracturing, however, is highly water intensive. The problems are made even worse by the fact that water resources are unevenly distributed spatially and temporally. Except for the Sichuan basin, the other major shale gas basins in China are located in the arid and water-scarce north and northwest. Therefore, the development of shale in China may be severely hampered by lack of water resources.

## 7.5   Environmental Issues

Hydraulic fracturing is a technology that involves injecting water, chemicals and proppants to break shale rock and release the gas locked in it. Hydraulic fracturing is controversial due to the environmental risks involved in the fracturing or 'fracking' process, which can result in issues of landscape and land usage, ground water contamination, minor earthquakes, waste water disposal and greenhouse gas emissions after fracturing.

*7.5.1   Land Consumption.* During the construction phase for a well pad and associated infrastructure, such as unimproved or gravel roads for trucks, water storage, gas processing and transporting facilities, *etc.,* land is needed. China has only 7% of the world's farmland and *per capita* farmland occupation is only about half of the world average rate. In the last several decades, remarkable economic success in China has caused rapid loss of farmland to residential, industrial and commercial uses. This further restricts access to land for large-scale shale gas drilling.

*7.5.2   Aquifer Water Contamination and Waste Water Disposal.* The contamination of aquifer water and wastewater from hydraulic fracturing has been one of the most prominent areas of debate. This is due, in part, to the success of the 2010 documentary, *Gas land,* which largely focused on the issue of water contamination from fracturing fluids containing hazardous substances such as benzene, toluene, formaldehyde or hydrochloric acid and methane. Fracturing fluids, methane and $H_2S$ from fracked shale formation in China could get into the aquifer water and cause serious contamination if a well is not properly cemented. Flow-back wastewater may contain salt, chemicals, heavy metals and radioactive materials, which could be problem if not properly processed.[15]

*7.5.3   Greenhouse Gas and $H_2S$ Emissions.* Compared with conventional gas, the development of shale gas has a higher risk of greenhouse gas (methane and $CO_2$) release. Some shale gas reserves in China also potentially contain hazardous levels ($>1\%$) of $H_2S$. Inappropriate collection and treatment of these air pollutants will affect the local air quality. Given the high population density in China, the impacts of air pollution and climate change should not be neglected.

*7.5.4 Earthquakes.* Many independent studies show that hydraulic fracturing can potentially lead to earthquakes; for the tectonically active and
historical earthquake areas, *e.g.* some areas in the Sichuan Basin, the
hydraulic fracturing could fracture pre-existing faults and cause ground
movements.

# 8 The Promising Future of China's Shale Gas Development

The largest shale gas reserve in China provides opportunities for exploration
and production, manufacturing, service, distribution, storage and downstream industries. Even though China is lacking mature technologies
to develop shale gas in complex geological settings, new technology is
rapidly becoming a worldwide commodity through the efforts of major
service companies. China and the shale experience-rich US share an interest
in the domestic and international development of shale gas resources
and US-China business-to-business partnerships, as well as government-to-
government cooperation, will boost shale gas development in China.

So far, the Chinese government has been implementing new policies and
has invested a lot of money in research and approved a pilot project for shale
gas development to attract international oil companies to conduct joint exploration in China. China is also investing in shale-gas development in the
US in order to learn the relevant technologies from its partners. At the same
time, through trial-and-error in the pilot shale-gas areas in Changning-
Weiyuan in Sichuan, Fuling in Chongqing and the Yanchang block in Ordos,
companies have improved their understanding of how to fracture shales in
China. With limited participation from established global service companies, such as Baker Hughes and Schlumberger, Sinopec's Jianghan oilfield
has improved knowledge in the key areas of fracturing and logging. At one
well in the Sichuan Basin, Sinopec-Jianghan performed a 22-stage hydraulic
fracturing and the test results showed the successful commercial flow of
shale gas from marine shale in the Jiaoshiba area of Fuling, Chongqing. The
horizontal well costs fell from 100 million CNY (Chinese currency: renminbi)
to 50 million CNY per day in the Jiaoye area. With the progress made in
large-scale drilling, completion and mass-production, and the re-use of
drilling and eco-friendly fracturing fluids, these shale gas development costs
should be reduced further. Sinopec has developed a non-toxic and biodegradable synthetic base drilling fluid as an alternative to oil- and water-
based drilling muds, and supercritical $CO_2$ and $N_2$ if used for fracturing high
clay content shale. Sinopec has further plans to drill in the SE Chongqing/
E Sichuan Basin and the NE Sichuan Basin, due to recently discovered
commercial shale gas from both marine and lacustrine shales. PetroChina is
focusing its efforts on the Changning Block and the Weiyuan Block within
the Sichuan Basin. Ning 201 and Ning 203 have produced a good gas show
tested in the Silurian Longmaxi and Cambrian Qiongzhusi formations in the
Changning Block. Wei 201-H1 and other wells are reported to have commercial gas in the Longmaxi formation from hydraulic fractured horizontal

wells in the Weiyuan Block. PetroChina plans to drill more than 100 horizontal shale gas development wells in the next two years in the Sichuan
Basin. These developments indicate that China should fulfill its shale gas
development plan in several years. In June 2013, CNPC announced that it
has started building the country's first dedicated shale-gas pipeline in
Sichuan province – more pipelines will be built with both government
support and private capital.

Recently, the new third plenary session of the 18th Communist Party of
China (CPC) Central Committee issued a detailed reform of the fiscal system
to loosen its grip on capital controls and allow private capital to play a
greater role in China's economy. Private and foreign investment will also be
welcomed to help develop shale gas in China. Despite environmentalists'
fears, the possibility of contamination is small since shale rocks are generally located much deeper than water aquifers and there are many non-
permeable rock layers between water aquifers and shale-rock formations.
China's shale gas is at least 2000 metres underground, *i.e.* significantly
deeper than shallow aquifers for drinking water and separated from
underground water by impermeable rock. Several thousand feet of cement
also seals off the annular space between the wellbore wall and the outside of
the well casing to prevent shale gas and formation fluids migrating up to the
freshwater aquifer. Even though a recent study shows that groundwater
contamination was caused by shale gas production in the US,[16] this was
found to be mainly due to poor well integrity or poor cementing. China can
use this information as a wake-up call to monitor the careful drilling and
stimulation of shale gas development.[17] Experiences learned from the US
will help China to extract shale gas effectively and environmentally by
implementing regulations. Indeed, China has scrutinised the well planning
and environmental protection plans; issued a ban on direct emissions of
waste gas; prescribed a more efficient use of water and energy, and a timely
rehabilitation of land, *etc.*

## 9   Conclusions

Shale gas production is an indispensable opportunity for China to obtain
energy security to power its booming economy and reduce air pollution.
Even though China's shales present challenges for hydraulic fracturing due
to their complex geological settings, sub-surface and ground conditions, as
well as the need for the development specialist technologies and a suitable
infrastructure, they still offer a very promising prospect. China's shales
represent the largest shale gas resources in the world, and low land costs,
relatively cheap labour, governmental support and policies, stronger environmental regulations and improving technologies favour their exploitation. Whilst shale gas developments in Poland have proven disappointing
and in France have faced a number of policy obstacles, shale gas developments in China have made promising progress with new breakthroughs in
indigenous and foreign technology, and the commercial production of shale

gas flow in the Sichuan Basin by PetroChina and Sinopec. A better under-standing of China's shale gas reservoirs, combined with more experience in horizontal drilling and hydraulic fracturing, should facilitate the con-struction of deeper wells with longer laterals for targeting the most pro-ductive interval of the shales. As shale gas takes off in China, this should help to secure China's energy supply to further power its economy, with social and environmental benefits, as well as providing the country with more leverage over geopolitical issues.

# References

1. P. Stevens, *The Shale Gas Revolution: Hype and Reality,"* A Chatham House Report, 2010.
2. S. Bierman, *U.S. Overtakes Russia as Biggest Natural Gas Producer (Up-dated)*, Bloomberg, January 12, 2010.
3. Economist Intelligence Unit, *A Greener Shade of Grey*, ed. M. Adams, 2011; http://country.eiu.com/article.aspx?articleid = 1809063365 (last accessed 12/11/2013).
4. F. Gao, *Will there be a Shale Gas Revolution in China by 2020*, The Oxford Institute for Energy Studies, University of Oxford, 2012, NG61, 1–41.
5. S. Lu, The Weekly Wrap: The four squabbling fiefdoms in China's shale-led political transition, *Foreign Policy: The Oil and the Glory*, 2012; http://oilandglory.foreignpolicy.com/posts/2012/06/28/the_weekly_wrap_june_29_2012_part_ii (last accessed 17/9/2013).
6. C. Zou, C. Xu, Z. Wang, S. Hu, G. Yang, J. Li, W. Yang and Y. Yang, Geological characteristics and forming conditions of the platform mar-gin large reef-shoal gas province in the Sichuan Basin, *Pet. Explor. Dev.*, 2011, **38**(6), 641–651.
7. J. Zhang, B. Xu, H. Nie, Z. Wang and T. Lin, Exploration potential of shale gas resources in China, *Nat. Gas Ind.*, 2008, **28**(6), 136–140 (in Chinese with English abstract) (in Chinese with English abstract).
8. IHS Global Insight, *The Economic and Employment Contributions of Shale Gas in the United States*, 2011; http://www.ihs.com/images/Shale_Gas_Economic_Impact_mar2012.pdf (last accessed 9/3/2013).
9. Z. Feng, *The Impact of the Changing Global Energy Map on Geopolitics of the World*, China-United States Exchange Foundation, 2013; http://www.chinausfocus.com/energy-environment/the-impact-of-the-changing-global-energy-map-on-geopolitics-of-the-world/ (last accessed 07/25/2013).
10. D. Zhang, Main solution ways to speed up shale gas exploration and development in China, *Nat. Gas Ind.*, 2011, **31**(5), 1–5 (in Chinese with English abstract) (in Chinese with English abstract).
11. R. Gold and C. Dawson, Chinese energy deals focus on North America: State-owned firms seek secure supplies, advanced technology, *Wall Street J.*, 2013; http://online.wsj.com/news/articles/SB1000142405270230

468250457915335037908908 2?KEYWORDS = China+oil (last accessed 19/11/2013).

12. National Development and Reform Commission of China, Ministry of Finance of China, Ministry of Land and Resources of China, Ministry of Commerce of China, *Shale Gas Development Plan (2011–2015)*, 2012; http:// zfxxgk.nea.gov.cn/auto86/201203/t20120316_1454.html (last accessed 20/ 03/2013) (in Chinese).

13. S. Izrailova, *Shale Gas Development in China*, 2013; http://www.e-ir.info/ 2013/01/09/shale-gas-development-in-china/ (last accessed 13/11/2013).

14. The World Bank, *Renewable Internal Freshwater Resources Per Capita (cubic meters)*, 2011; http://data.worldbank.org/indicator/ER.H2O.INTR. PC (last accessed 01/12/2013).

15. A. Vengosh, N. Warner, R. Jackson and T. Darrah, The effects of shale gas exploration and hydraulic fracturing on the quality of water resources in the United States, *Procedia Earth Planet. Sci.*, 2013, 7, 863–866.

16. J. Tollefson, Gas drilling taints groundwater, *Nature*, 2013, **498**, 415–416.

17. H. Yang, R. J. Flower and J. R. Thompson, Shale gas: pollution fears in China, *Nature*, 2013, **499**, 154–154.

# Unconventional and Unburnable: Why going all out for Shale Gas is the Wrong Direction for the UK's Energy Policy

TONY BOSWORTH

ABSTRACT

Despite the Prime Minister's statement that we must go 'all out for shale', there remain major doubts about whether this is the right direction for the UK's energy policy. Unconventional oil and gas, such as shale gas, risk locking the UK into the use of fossil fuels at a time when meeting climate change targets means we need to be reducing their use rapidly. Despite claims by the industry and its supporters that fracking is safe, there is considerable evidence of impacts on the local environment and human health from the US, where unconventional gas production has grown significantly in the last decade, and from Australia. There are also doubts about whether the UK's proposed regulation of the industry will be adequate. The Prime Minster claims that shale gas will cut UK energy bills, despite most experts thinking that this is, at best, unlikely. The UK needs to move fast towards an energy system based on much-improved energy efficiency and using the UK's huge potential for renewables. In this context, shale gas is a gamble we do not need to take.

## 1 Introduction

The UK Government has nailed its colours firmly to the shale gas mast. The Prime Minister, David Cameron, has said that we must go "all out for

Issues in Environmental Science and Technology, 39
Fracking
Edited by R.E. Hester and R.M. Harrison
© The Royal Society of Chemistry 2015
Published by the Royal Society of Chemistry, www.rsc.org

shale".[1] But despite this gung-ho enthusiasm, there remain major doubts as to whether it is the right direction for UK energy policy.

Friends of the Earth strongly believes that unconventional gas and oil (such as shale gas) are not the answer to the UK's energy problems because:

- It risks keeping the UK locked in to the use of gas at a time when we need to reduce fossil fuel use significantly in order to play our part in tackling climate change;
- Evidence from the United States, where fracking has grown rapidly in the last decade, and elsewhere shows that the process of fracking involves risks to the local environment and to human health; and
- Despite Government claims to the contrary, most experts think UK shale gas is unlikely to cut energy bills.

In our opinion, shale gas – and other unconventional gas and oil – is both unconventional and unburnable. This chapter explores these issues and looks at the real solutions to the UK's energy problems.

## 1.1 The Government's View

The UK Government sees shale gas as a vital part of the UK's energy future. The Secretary of State for the Environment, Food & Rural Affairs, Owen Paterson, has described the UK's shale gas potential as "an extraordinary ... windfall, which is entirely God-given".[2]

The Government is determined, in the words of the Chancellor of the Exchequer, George Osborne, in his 2013 Budget speech, to "make it happen". To this end, Mr Osborne is introducing tax breaks for shale gas production as part of a fiscal regime which he wants to make the most generous in the world. His Cabinet colleague Eric Pickles, the Secretary of State for Communities & Local Government, has relaxed planning guidelines to boost the shale gas industry. For example, drilling companies no longer have to inform householders directly if drilling and fracking might take place below their property. The UK Government has actively manoeuvred against further regulation at the EU level because this would delay the development of the industry in the UK. This is despite the fact that the Impact Assessment produced by the European Commission showed that a Directive setting specific requirements for the exploration and production of unconventional oil and gas would be much more effective in reducing environmental impacts and risks, providing legal clarity and addressing public concerns, than the preferred option (and that supported by the UK) of a non-binding recommendation.[3]

## 2 Environmental Concerns

There are sharply conflicting views on the scale of the risks and the likelihood of them occurring.

Shale gas advocates claim that fracking is safe, with no local environmental risks. Ed Davey, the UK Secretary of State for Energy & Climate Change, drawing on a report from the Royal Society and the Royal Academy of Engineering, has said that fracking "can take place quite safely, without hurting the local environment. It will not contaminate water supplies".[4]

But a report for the European Commission, *Support to the Identification of Potential Risks for the Environment and Human Health arising from Hydrocarbons Operations involving Hydraulic Fracturing in Europe*,[5] identified "a number of issues as presenting a high risk for people and the environment". It assessed the cumulative impact of fracking at multiple installations as "high risk" in terms of groundwater contamination, surface water contamination, water resources, release to air, land take, risk to biodiversity, noise impacts and traffic. In a Global Environmental Alert issued in 2012, the United Nations Environment Program concluded that fracking "may result in unavoidable environmental impacts even if (unconventional gas) is extracted properly, and more so if done inadequately".[6] In 2013, drilling company Cuadrilla was censured by the Advertising Standards Authority for writing in a public information leaflet that it used "proven, safe technologies".[7]

The United Nations Environment Program lists the following potential risks from shale gas exploration and production:

- Greenhouse gas emissions;
- Land footprint (on natural and crop land);
- Risk of explosion (wells, pipelines, transport);
- Risk of leakage from wells into the water table;
- Risk of leakage from fracking fluid or from produced water into the water table;
- Risk of leakage from improperly treated produced water and fracking fluids from flowback into the soil and water table and surface water;
- Risk of infiltration of fracking fluid into the water table;
- Risk of migration of naturally occurring toxic substances; and
- Impacts from improperly treated water in crops.[6]

Friends of the Earth believes there are real concerns about local and global environmental impacts. This chapter cannot consider all of these potential impacts in detail but will look briefly at some of them.

## 2.1   Climate Change

Exploiting shale gas poses major risks for climate change by perpetuating our use of fossil fuels rather than moving to low-carbon alternatives. There are three key issues to consider:

- The comparative emissions of shale gas compared to conventional gas and other fossil fuels;
- The impact on overall emissions of using shale gas; and
- The impact on investment in renewable energy.

Broderick and Wood's chapter in this book (see Chapter 4) considers the comparative emissions of shale gas and other fossil fuels, so this chapter will look at the other two issues.

Climate change is arguably the greatest existential threat that the world currently faces. The most recent report from the Intergovernmental Panel on Climate Change (IPCC), released in 2013,[8] found that warming of the climate system is unequivocal and, since the 1950s, many of the observed changes are unprecedented over decades-to-millennia. The atmosphere and oceans have warmed, the amounts of snow and ice have diminished, sea level has risen and the concentrations of greenhouse gases (GHGs) have increased. It is extremely likely that human influence has been the dominant cause of the observed warming since the mid-20th century. This warming has already increased the occurrence of some types of extreme weather events. Further temperature increases will see the frequency of heavy rainfall events and very hot days increase, with the severity increasing in proportion to warming. In some regions there will be more droughts and floods, though a lack of data prevents any firm conclusions on how the frequency will change globally.

Governments of developed countries, including the UK, have repeatedly stated that, to avoid dangerous climate change, global temperatures must rise by no more than 2 °C above pre-industrial levels. For example, the EU's target is explicit that "The EU must adopt the necessary domestic measures and take the lead internationally to ensure that global average temperature increases do not exceed pre-industrial levels by more than 2 °C".[9]

Challenging though this is, it might not be enough. Since the 2 °C target was first adopted, knowledge about climate change and its potential impacts has progressed and many scientists now believe that potentially catastrophic impacts will be felt at lower temperature rises. This is demonstrated by the 'Burning Embers' diagram, first produced by the Intergovernmental Panel on Climate Change (IPCC) in 2001.[10] This shows the likelihood of various types of impact of climate change (risk of extreme weather events, risk of large-scale discontinuities) for different global temperature rises. The updated version of this diagram, produced in 2009, makes it clear that for a 2 °C temperature rise there are now substantial negative impacts associated with extreme weather events and that there are now "moderately significant" risks of large scale discontinuities – risks which were thought to be "very low" in 2001.[11] Such evidence has led over 100 developing-country Governments to call for a maximum temperature rise of no more than 1.5 °C above pre-industrial levels. The 'Burning Embers' diagram will be updated by the IPCC in March 2014.

However, even accepting the 2 °C target, the carbon maths is clear: if we are to keep the likely global temperature rise below 2 °C, we cannot afford to burn unabated more than a small fraction of the world's current proven fossil fuel reserves, let alone any as yet unproven reserves. Exploring for new hydrocarbon resources is simply adding to the stock of already unburnable carbon. The concept of 'unburnable carbon' started to gain traction in 2013 following important work by the Carbon Tracker Initiative.[12] In the words of

Professor Dieter Helm of Oxford University "There is enough oil and gas (and coal too) to fry the planet several times over".[13]

Shell has responded to Carbon Tracker's work, saying "There is really nothing to argue about in terms of the $CO_2$ math itself. It is certainly the case that current proven reserves will take us well past 2 °C if completely consumed and the $CO_2$ emitted".[14]

However, the UK Government's policy is to maximise hydrocarbon recovery. In June 2013, Energy Secretary Ed Davey wrote that "maximising recovery of the UK's indigenous supplies of oil and gas" was important economically and for the UK's energy security.[15]

Maximising hydrocarbon recovery is not compatible with the UK playing its part in meeting the global 2 °C climate change target which the UK Government has repeatedly endorsed. The UK Government needs to leave some of its proven or potential fossil fuel resources in the ground and, given the risk of unavoidable and serious environmental impacts associated with its extraction and the uncertainty over the scale of resources, the starting point for this should be unconventional gas, including shale gas.

The background to this is explained below and is drawn from a Friends of the Earth briefing paper *The UK, Shale Gas and Unburnable Carbon: Questions for the Government.*[16]

*2.1.1 Carbon Budgets and Unconventional Gas.* The work of Meinshausen *et al.* on carbon budgets has assessed the percentage chance of exceeding a given temperature target for a given amount of global carbon dioxide emissions.[17] This shows that if we are prepared to accept a 50 : 50 chance of the global temperature rise exceeding 2 °C, then the world can emit 1400 Gt $CO_2$[18] between 2000 and 2049.[†] However, if we want only a 1 in 3 chance of a greater than 2 °C global temperature rise, then the global carbon budget for 2000–2049 falls to 1150 Gt $CO_2$.

Clearly, some of this carbon budget has already been used up since 2000. Subtracting global carbon emissions for 2000–2012 gives the remaining global carbon budget for 2013–2049.[18] For a 50 : 50 chance of going over two degrees, this is 1050 Gt $CO_2$ and for keeping to a 1 in 3 chance of going over 2 °C, the budget is 750 Gt $CO_2$. The IPCC's 5th Assessment Report, released in September 2013, gives very similar carbon budgets.

The IPCC has definitions of different likelihoods, ranging from 'virtually certain' (a 99–100% likelihood) to 'exceptionally unlikely' (a 0–1% likelihood). Given the clear stating of the EU's target (see previous), Friends of the Earth believes that the highest level of risk compatible with this 'do not exceed' wording would be to keep the likelihood to 'unlikely' or lower. This translates to a less than 33% chance of exceeding 2 °C. (The UK's binding climate change target – the first in the world - implies only a 50 : 50 chance of staying below 2 °C). This implies that the appropriate remaining global carbon budget would be 750 Gt $CO_2$, though even this accepts a relatively

---

[†]All figures are rounded to the nearest 50 – for precise numbers, see Ref. 16.

**Table 1**   Global fossil fuel reserves.

|        | Proven reserves          | Emissions from proven reserves (Gt $CO_2$) |
|--------|--------------------------|-------------------------------------------|
| Oil    | 236 billion tonnes       | 724                                       |
| Gas    | 187 trillion cubic metres | 396                                      |
| Coal   | 861 billion tonnes       | 1667                                      |
| Total  |                          | 2787                                      |

high level of risk for something that we have agreed we must avoid (by comparison, we insure our homes for levels of risk of less than 1%) and a temperature target above that advocated by over 100 countries.

However, proven global fossil fuel reserves, if burned, would emit 2800 Gt $CO_2$ see Table 1).[19] (So burning the world's proven fossil fuel reserves would emit more than three times the emissions associated with a safe global carbon budget).

This does not take account of unconventional reserves. The International Energy Agency (IEA) has developed a 'Golden Age of Gas' scenario.[20] This assumes "an accelerated global expansion of gas supply from un-conventional resources" so that "demand for gas grows by more than 50% to 2035" which "puts $CO_2$ emissions on a long-term trajectory consistent with a probable temperature rise of more than 3.5 degrees Celsius (°C) in the long term, well above the widely accepted 2 °C target". The IEA's Chief Economist, Fatih Birol, has admitted "we are not saying that it will be a golden age for humanity – we are saying it will be a golden age for gas".[21]

*2.1.2   Implications for the UK.*   If the world can only burn approximately a third of its proven fossil fuel reserves, then what does this mean for individual countries? Climate change is a global problem, requiring the co-operation of all countries. The Framework Convention on Climate Change, signed by 195 nations, states that "The Parties should protect the climate system for the benefit of present and future generations of humankind, on the basis of equity and in accordance with their common but differentiated responsibilities and respective capabilities. Accordingly, the developed country Parties should take the lead in combating climate change and the adverse effects thereof".[22]

The issues of responsibility, fairness and justice run through international climate negotiations. The global temperature rise is due to a combination of both future and historic emissions: the cumulative build-up of greenhouse gases in the atmosphere. Put simply, developed countries are over-whelmingly responsible for historic emissions, and so developing countries have a greater claim to remaining future emissions. As a result, splitting any remaining global carbon budget by a country's share of the world population would be equal, but not compatible with the Convention's or the subsequent Copenhagen Accord's "differentiated responsibilities" – for it would be ignoring all historic responsibility.

It is hard to see why any developing country would accept any limits on its own production if countries which have used vast amounts of fossil fuels in the past are able to ignore this historical responsibility. This means the 'start date' for calculations of national carbon budgets needs to be amended. A start date of 1990 – widely used in climate negotiations – seems an absolute minimum. Taking into account historical emissions more comprehensively would mean a much earlier start date.

Adding the actual global carbon emissions for 1990–1999 (234 Gt $CO_2$) to Meinshausen's 2000–2049 global carbon budget for a 33% chance of not exceeding a 2 °C global temperature rise (1158 Gt $CO_2$)[17] gives a global figure of 1400 Gt $CO_2$ (rounded to the nearest 50 Gt). On a *per capita* basis, the UK's share of this budget is 12.2 Gt $CO_2$.

How much of this have we used already? Since 1990, the UK has extracted oil, gas and coal resulting in emissions of 12.1 Gt $CO_2$. On this basis, the UK's remaining carbon budget is 0.1 Gt $CO_2$. The use of production rather than consumption figures here is deliberate. Consumption levels are clearly critical, but production and consumption are linked, and production in one country can affect consumption in another: shale gas production in the US has led to coal exports to Europe (see section 2.1.3).

Maintaining and indeed increasing fossil fuel production creates additional pressure to maintain fossil-fuel-based consumption patterns and makes it harder to reduce demand. Global carbon emissions need to peak and decline almost immediately – this requires a twin approach of reducing demand and reducing supply of fossil fuels. Friends of the Earth believes that the prospects of a strong global deal on climate change are made much less likely if developed countries like the UK continue to accelerate their own fossil fuel production programmes.

However, the UK is still aiming to maximise hydrocarbon recovery. The Government's central projections for conventional fossil fuel production would emit 3.7 Gt $CO_2$ – already many (37) times our remaining carbon budget. The production of unconventional gas and oil would add significantly to this, including a potential additional 7 Gt $CO_2$ from shale gas, based on the recovery of 130 trillion cubic feet (tcf) of shale gas, which is 10% of the central (P50) total gas-in-place estimate in the Bowland Shale.[23]

The potential for shale gas is underpinning the Government's desire to build new gas-fired power stations, risking lock-in to gas. The Government's official advisors, the Committee on Climate Change, have written of the dangers of such a 'dash for gas': "The apparently ambivalent position of the Government about whether it is trying to build a low-carbon or a gas-based power system weakens the signal provided by carbon budgets to investors (and is) damaging prospects for required low-carbon investments".[24] Similar views have been expressed by the respected think-tank Chatham House: "also significant is the argument that a ramping up of shale gas operations could come at the expense of needed investment in renewable energies like wind and solar".[25]

The Tyndall Centre has calculated that the cost of building shale-gas-generated electricity capacity in the UK (well costs plus power station costs) could provide the same electricity generation capacity from onshore and offshore wind.[26]

*2.1.3 'Instead of' or 'As well as'?*. Fatih Birol of the IEA has said that an increase in unconventional gas production would be "a good move if it replaces coal... but it is definitely not the optimum path".[27] This comment highlights another global-level issue: can we ensure that shale gas will be used instead of other fossil fuels, such as coal or conventional gas from the Arctic, rather than as well as them?

This has not been the case in recent years in the US. Analysis by the Tyndall Centre for Climate Change Research shows that fuel switching (from coal to shale gas) in the US power sector may account for up to half the total national emissions reduction.[28] However, the consequent drop in the price of coal has led to an increase in exports, including to the UK where coal use for electricity generation in the UK rose from 30% in 2011 to 39% in 2012, the latest full year for which figures are available.[29] Tyndall estimates that just over half of the emissions avoided in the US power sector have been 'exported' as coal.

*2.1.4 Carbon Capture and Storage.* Many fossil fuel advocates claim that problems of carbon emissions can be addressed through the use of carbon capture and storage (CCS). CCS involves the capture of carbon dioxide emissions from fossil-fuel burning and their storage underground where they cannot have an impact on the atmosphere (see *Carbon Capture*, vol. 29, in this Issues in Environmental Science series).

Friends of the Earth believes that CCS may have a transitional role to play in future energy policy, retrofitting many current gas-fired power stations in the UK and elsewhere, and also being used in industry (*e.g.* cement, steel) but it is no panacea:

- It can only deal with some fossil-fuel emissions – on power stations and large industrial processes, but not on domestic and commercial heating which accounts for much of the UK's gas consumption;
- It does not eliminate carbon emissions completely;
- The use together of the three elements of CCS (capture, transport and storage) is still commercially unproven at commercial scale; and
- It does not have a large impact on carbon budgets. Carbon Tracker has estimated that even an "idealised" scenario of huge investment in CCS infrastructure would only extend global carbon budgets by 125 Gt $CO_2$.[30]

At present, CCS seems to be more a justification for allowing coal and gas investment to go ahead on the speculative assumption that CCS will be fitted in the future.

In conclusion with regard to climate change, extracting shale gas is just adding to the stock of unburnable carbon at both the global and UK levels. It is underpinning the Government's 'dash for gas' and could adversely affect investment in renewables.

## 2.2 Water Resources

Fracking is a water-intensive process. Estimates of water use vary from 250 – 4000 m$^3$ for drilling and 7000–23 000 m$^3$ for fracking, depending on factors such as geology, depth and thickness of the shale.[31]

In the UK, the total volume of water that could be needed if a fracking industry were to develop has been assessed as relatively low at the national level in comparison to other uses. However, there could be local and regional concerns. The Weald in South East England is a key target area for un-conventional fossil fuels but the water resources in many parts of the region are already over-abstracted. The pressures of increased demand (from population growth, a growth in the number of households and economic growth), reform of the abstraction regime and the impacts of climate change on weather patterns could add to these problems. Water UK has commented that "where water is in short supply there may not be enough available from public water supplies or the environment to meet the requirements for hydraulic fracturing".[32]

## 2.3 Water Contamination

The risk of water contamination is one of the most controversial issues in the fracking debate. The industry and its advocates point to studies such as 'Fact-Based Regulation for Environmental Protection in Shale Gas Develop-ment' from the Energy Institute of the University of Texas at Austin,[33] which concluded that "there is at present little or no evidence of groundwater contamination from hydraulic fracturing of shales at normal depths" and make claims such as "Hydraulic fracturing has been in use for more than 60 years without any confirmed cases of groundwater contamination".[34] However, it is necessary to examine such claims more closely.

High-volume hydraulic fracturing, such as is carried out today in the US and as is proposed for the UK, has not been in use for more than 60 years. The first well fracked was in 1947, but this was a vertical well using 1000 gallons of napalm injected at relatively low pressure. High volume hydraulic fracturing has been made possible by technical developments in the last decade such as horizontal drilling, the use of friction-reducers to create 'slickwater' and drilling several wells from one well-pad.

The claim that there have been no proven cases of groundwater con-tamination is also questionable. The US Environmental Protection Agency announced that it did not intend to complete an investigation into the pollution of an aquifer in the town of Pavillion, Wyoming, although it had previously stated in a draft report that "the data indicates likely impact to ground water that can be explained by hydraulic fracturing".[35] Some claim that this is not the only case where the EPA has stopped its investigations.[36] It is also suspected that many claims about groundwater contamination have been settled out of court with confidentiality agreements.[37]

Finally, the industry and its advocates typically adopt a very narrow def-inition of the word fracking, referring just to the stimulation of the well. This leads to claims that problems caused by well failure or accidents involving trucks carrying fracking chemicals are not ascribed to fracking.[38]

One study of aquifers overlying the Marcellus and Utica shales in the north-eastern US found "systematic evidence of methane contamination of groundwater associated with shale gas extraction".[39] A review by Associated Press (AP) of complaints about well-water contamination in Ohio, Pennsylvania, Texas and West Virginia found many confirmed cases of water pollution from oil and gas drilling. AP concluded that the review "casts doubt on industry suggestions that such problems rarely happen".[40]

Methane and fracking fluid may escape/contaminate water *via* a number of different routes:

- Migration down naturally occurring fractures in the rock or *via* extension of fractures created by fracking or *via* nearby abandoned wells;
- Leaks *via* well-casings that have been inadequately completed or which have subsequently failed (probably a greater risk than that above); and
- Leaks or spills of fracking fluid or 'produced water' above ground.

If there is a risk of contamination, what chemicals could be involved? A major problem is that there is limited data on the chemicals that have been used for fracking. This is because US law excludes fracking from federal regulation by the Environmental Protection Agency. However disclosure is required by some US states and some companies are posting the composition of the fracking fluid they are using online. In the UK, companies will be required to publish the contents of fracking fluid.

One study in the US looked at 944 products containing 632 chemicals used in natural gas operations and assessed the health impacts of 353 chemicals. According to the authors "more than 75% of the chemicals could affect the skin, eyes, and other sensory organs, and the respiratory and gastrointestinal systems. Approximately 40–50% could affect the brain/nervous system, immune and cardiovascular systems, and the kidneys; 37% could affect the endocrine system; and 25% could cause cancer and mutations".[41]

A study reported in the *Journal of Endocrinology* analysed water samples taken from sites near active fracking wells in Garfield County, Colorado, and from control sites in the same and an adjoining county.[42] The study found that the samples from sites near active fracking wells contained higher levels of endocrine-disrupting chemicals that could increase the risk of reproductive, neurological and other diseases.

The fracking industry claims, rightly, that the chemicals used are only a very small percentage of the total volume of liquid that is pumped underground. However, given the volume of water used in fracking, this still represents a very large quantity of chemicals. Fracking one well uses between 9 and 29 million litres of water, so even if chemicals account for only 0.5% of the water volume, this means between 45 000 and 145 000 litres of chemicals used per well.

In addition to the chemicals, fracking waste water may also contain substances from deep underground such as strontium, benzene, toluene and Naturally Occurring Radioactive Material (NORM) such as Radium 226.

## 2.4   Health Impacts

In addition to potential problems caused by water contamination, fracking for shale gas has also been linked to increased levels of air pollution and associated health problems. Air pollution was identified as a 'high risk' by the European Commission study mentioned above.

In 2013 Public Health England (PHE), an executive agency of the Department of Health, produced a draft report *Review of the Potential Public Health Impacts of Exposures to Chemical and Radioactive Pollutants as a Result of Shale Gas Extraction*. The general conclusion was that "the potential risks to public health from exposure to the emissions associated with shale gas extraction are low if the operations are properly run and regulated".[43] With regard to air pollution, PHE concluded that "the available evidence suggests that while emissions from individual well pads are low and unlikely to have an impact on local air quality, the cumulative impact of a number of well pads may be locally and regionally quite significant".

An initial assessment by McKenzie *et al.*[44] looking at birth outcomes found an association between the density and proximity of natural gas wells within a 10-mile radius of a mother's home and the prevalence of congenital heart defects and neural tube defects (defects of the brain, spine or spinal cord). The authors stated, however, that more detail was needed on levels of exposure during pregnancy.

A separate assessment by McKenzie *et al.*[45] looked at the risks from exposure to hydrocarbons, notably benzene, near fracking wells in Colorado. It found that the risk of cancer was increased for people living within half a mile of a well compared to those living more than half a mile away – from 6 in a million to 10 in a million. It also found that non-cancer health impacts were greater for those living nearer wells.

PHE indicated that their assessment was in some cases based on limited evidence – for example, at the time of their report, there had only been one peer-reviewed assessment of the health impacts of air emissions from shale gas operations: the 2012 McKenzie study mentioned previously. However, as the US group, Physicians, Scientists and Engineers for Healthy Energy (PSE), has written, with reference to the health impacts of fracking, "lack of data is not an indication of an absence of harm".[46] According to PSE, history offers several examples of where "health-damaging industrial activities were scaled much more rapidly than the science of its health effects and subsequent evidence-based policy development", including tobacco, PCBs, asbestos and leaded petrol. PSE conclude that "the science should be put before risky industrial processes are allowed to be scaled".

In the absence of adequate data, we believe the Precautionary Principle should apply. Friends of the Earth agrees with the authors of a paper in the

*American Journal of Public Health* that "The paucity of scientific evidence looking at the public health impact of natural gas extraction complicates the issue. It is difficult and potentially dangerous to formulate policy and regulations in a vacuum".[47]

## 2.5 Earthquakes

The rise in concern about fracking in the UK started when test-fracking by Cuadrilla in Lancashire triggered low-level earthquakes, although it says that these tremors were far too weak to cause any damage to buildings or harm to people.[48]

However, a possibly even greater risk is to the integrity of the well-casings, typically made of steel and cemented in place, designed to reduce or eliminate the possibility of leaking methane or flowback water. Tony Grayling, Head of Climate Change at the Environment Agency, has acknowledged that this could be a problem. Referring to the Lancashire earthquakes, which did damage the well-casing,[49] he said, "we need to understand what is the maximum damage that might be done in such circumstances to a well and the integrity of the casing, whether it would increase the risk of a leak. If there is ground water in the vicinity, that could be a problem".[50]

Evidence from the US shows that the Lancashire experience is not unique. Several US states have experienced seismic activity following shale gas drilling and fracking in areas where this has not previously happened. Although a clear causal link has not been proved in all cases, a close correlation between fracking and earth tremors can be seen.[51]

## 2.6 Will UK Regulation be Adequate?

The Government and industry claim that the UK's regulatory regime is strong enough to prevent any problems. Typical of such views is that of the UK Onshore Operators Group: "The UK has much stricter regulations than anywhere else in the world and is recognised as the gold standard".[52] However, Friends of the Earth believes that better regulation can make the industry safer, but not safe. As has been referred to above, the United Nations Environment Programme has concluded "Hydrologic fracking may result in unavoidable environmental impacts even if (unconventional gas) is extracted properly, and more so if done inadequately. Even if risk can be reduced theoretically, in practice many accidents from leaky or malfunctioning equipment as well as from bad practices are regularly occurring".[6]

Friends of the Earth is concerned that the proposed regulation of unconventional gas and oil activity is flawed because:

- It fails to ensure that a precautionary approach is adopted;
- It fails to address the serious climate change impacts of burning unconventional gas and oil;

- It is inadequately enforced;
- It is not bespoke, despite the United Nations Environment Program, referencing the European Commission, stating that unconventional gas and oil operations "will require dedicated regulations".[6]


## 3  Local Economic Impacts

At the heart of the UK Government's support for shale gas is its claimed boost to the economy through lowering energy bills and creating jobs. In neither case is the issue as clear cut as the Government makes out.

### 3.1  Energy Bills

Rising energy prices are a key political issue in the UK. Household energy bills rose by 44% in real terms between 2002 and 2011.[53] The main reason for this was rising gas prices, which accounted for almost two-thirds of the increase in the average dual-fuel energy bill between 2004 and 2010.[54]

Claims about lowering energy bills have been key for the Government. The Chancellor of the Exchequer signalled his enthusiasm for shale gas when he told the Conservative Party Conference in 2012 that he wanted to make sure that Britain was not "left behind as gas prices tumble on the other side of the Atlantic". More recently, Prime Minister David Cameron has written that "fracking has real potential to drive energy bills down... gas and electric bills can go down when our home-grown energy supply goes up".[55]

But these claims have been widely refuted:

- The Secretary of State for Energy & Climate Change, Ed Davey, has said that "North Sea Gas didn't significantly move UK prices – so we can't expect UK shale production alone to have any effect";[56]
- Lord Stern, one of the world's leading climate-change economists, has described the Prime Minister's claims as "baseless economics";[57] and
- Lord Browne, the Chairman of Cuadrilla Resources (the leading drilling company in the UK) said that fracking in the UK "is not going to have material impact on price".[58]

Shale gas production costs in Europe are likely to be higher than in the US. Reasons for this include less-promising geology, higher population density and associated problems of land availability, the lack of a competitive on-shore drilling and services industry and tougher environmental regulation.[59] Factors such as these led the International Energy Agency (IEA) to conclude that operating costs in Europe will be 30–50% higher than in the US.[60]

Claims of cheaper gas prices also ignore global market dynamics. Demand for gas is rising fast, particularly from China, India and other emerging economies. This growing demand is likely to soak up new gas supplies, potentially keeping supply constrained and prices high, meaning that "UK households and industry would be tied to a highly unpredictable roller coaster of gas prices that are generally high and can spike higher due to volatility".[61]

Bloomberg New Energy Finance has estimated that "the cost of shale gas extraction in the UK will be between $7.10 and $12.20/MMBtu... similar to the range of market prices for natural gas seen in the UK during the course of 2012" and concluded that hopes that shale gas will lead to lower energy prices for the UK "should be treated as wishful thinking".[62] The IEA's analysis shows that gas prices in Europe will be around 40% higher than today in both 2020 and 2035.[61]

## 3.2 Jobs and Local Economy

The shale gas industry paints an overwhelmingly positive picture of its local economic impact through job creation. A report for Cuadrilla has claimed that shale gas production in Lancashire could create up to 6500 full-time equivalent jobs in the UK as a whole, with 1700 of these in Lancashire.[63]

The Institute of Directors has calculated that the production phase of a UK gas industry could support up to 74 000 jobs (direct, indirect and induced).[64] However, US experience shows that such claims should be treated with scepticism: numbers can be over-stated;[65] most employment is in the drilling phase, which only lasts around a year;[66] and many jobs go to transient workers who move from one well to another, with 70% of gas-well drilling jobs in Pennsylvania going to people from out of state.[67]

Job projection claims for UK shale gas vary greatly. The Environmental Report prepared for the Government ahead of a possible new onshore oil and gas licensing round assessed that the new licensing round could create between 2500 and 5000 jobs (direct, indirect and induced) in the low-activity scenario and 16 000–32 000 jobs in the high-activity scenario.[68]

Studies of job creation rarely look at potential negative impacts on other sectors. For example, the Regeneris research for Cuadrilla made no assessment of the possible impacts of fracking on tourism and agriculture. These are key parts of the local economy in the area where Cuadrilla hopes to drill.

## 4 Energy Security

Another key argument supporting shale gas in the Government's armoury is that it will improve the UK's energy security, providing a secure domestic source of gas to replace declining production from the North Sea, and reduce our dependency on imports.

However, there is no guarantee that any shale gas drilled in the UK will be used here. The UK is part of a wider European energy market and gas could be sold elsewhere in Europe.

What level of shale gas extraction would be needed in the UK to eliminate imports? Bloomberg New Energy Finance has calculated that this would require the drilling of 10 000–20 000 wells over a 15-year period, with a peak drilling level of around 1000 wells per year.[69] At the European level, analysis for the European Commission shows that, even under the most optimistic scenario for the development of shale gas in the EU, production only offsets the decline in conventional production and import dependence will stay at around 60% in 2040.[70]

However, this analysis looks at energy security purely from the perspective of security of supply. Friends of the Earth believes energy security should be defined more broadly than just about supply and geopolitics, and should also include price security (providing energy at reasonable prices) and environmental security (achieving emissions targets and minimising other impacts). In a report for Friends of the Earth, energy security expert, Professor Michael Bradshaw, concluded that "the best way to reduce the energy security risks associated with the UK's growing gas import dependence is to hold the course, promote renewable power generation, improve energy efficiency and reduce overall energy demand".[71]

## 5   Other Forms of Unconventional Gas

Much of the attention in the UK and elsewhere in Europe in recent years has focused on fracking for shale gas. But this is not the only form of unconventional fossil fuel that is arousing interest in the UK and Europe today. Companies are also investigating the potential for shale oil, coal bed methane and underground coal gasification.

Coal bed methane (CBM) is gas which is trapped in coal seams. To extract CBM, companies must 'de-water' the coal seam by drilling vertically and then horizontally (for up to 1 kilometre) and pumping out vast quantities of water. This releases the pressure in the coal seams and allows the gas to flow. CBM extraction does not always involve fracking – at least not in the early years of an operation. But if seams are not porous, or when gas flow starts to decline after a few years, wells are often fracked to increase productivity. In Australia, where CBM (known there as coal seam gas) is more developed, the industry estimates that up to 40% of wells end up being fracked[72] (see also Chapter 6 by Prof Alan Randall in this book).

Coal bed methane extraction brings many of the same problems as shale gas. Although CBM extraction does not always involve fracking, the chemicals used in CBM drilling muds can be just as toxic as those used in fracking, and there are the same risks of spills and leakages. And because CBM is typically found at much shallower depths than shale gas (up to 1200 m underground for CBM, compared to 2000–3000 m for shale gas), risks such as groundwater contamination are increased.

CBM waste water is extremely salty and has been found to contain not only harmful chemicals from the drilling fluids used, but also highly toxic BTEX (benzene, toluene, ethylbenzene and xylenes) chemicals, including known carcinogens, and naturally-occurring radioactive materials.[73] Spills and leaks of drilling fluids can also contaminate agricultural land and harm livestock. Vast quantities of contaminated water must be treated and disposed of. Evidence is emerging from Australia that existing treatments cannot remove all the toxins found in CBM wastewater.[74]

Communities living near CBM extraction sites in Australia have complained of respiratory problems, rashes and irritated eyes,[75] though research has not been done to corroborate a link. The risks posed to the local environment and human health have contributed to the imposition of a ban in New South Wales on CBM extraction within 2 km of residential areas.[76]

Underground coal gasification (UCG) involves the partial combustion underground of coal and the capture of the produced gas. By making it theoretically possible to use previously inaccessible coal seams, UCG can increase usable coal reserves – by 300–400% according to some sources. However, it involves both local environmental risks, notably subsidence and water contamination (one pilot project in Australia was shut down following the discovery of benzene and toluene in nearby groundwater monitoring bores),[77] and also contributes to climate change. This could be addressed by the use of CCS technology, and in the UK the Environment Agency has said this will be required if UCG is to be used for power generation.[78] However, one comprehensive analysis of the climate impacts of UCG has found that UCG is unlikely to be deployed widely with more than partial CCS, meaning that carbon emissions could still be significant.[79]

## 6   Friends of the Earth's Vision

Friends of the Earth believes that the priority for the UK's energy system is reducing our reliance on fossil fuels, basing our energy future on reducing energy waste and exploiting the UK's vast potential for renewables.

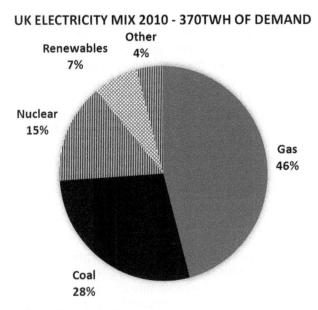

**UK ELECTRICITY MIX 2010 - 370TWH OF DEMAND**

Other 4%
Renewables 7%
Nuclear 15%
Gas 46%
Coal 28%

**Figure 1**   UK electricity mix 2010.

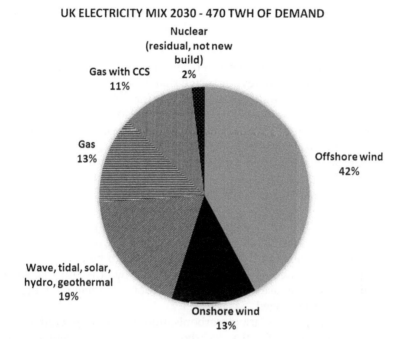

UK ELECTRICITY MIX 2030 - 470 TWH OF DEMAND

Nuclear (residual, not new build) 2%

Gas with CCS 11%

Gas 13%

Offshore wind 42%

Wave, tidal, solar, hydro, geothermal 19%

Onshore wind 13%

**Figure 2**   Possible UK electricity mix 2030.

## 6.1   The Power Sector

Using the *2050 Pathways* model developed by the Department for Energy & Climate Change, Friends of the Earth has shown how the UK can move from 75% of its electricity being generated from fossil fuels in 2010 to roughly 75% being generated from renewable sources in 2030 (see Figures 1 and 2).[80] This does not involve new nuclear power stations. By 2030, the UK relies on a mix of technologies: offshore wind, onshore wind, gas with CCS, unabated gas, solar, wave, tidal, geothermal and hydro power, coupled with greater levels of interconnection, energy storage and smart demand-management methods to ensure reliability and security of supply. Gas is still used, but its role is reduced.

The Committee on Climate Change (CCC) has set a target of a 60% cut in emissions by 2030 as a milestone on the route to the longer-term target of an 80% emissions cut by 2050. The CCC has concluded that, to meet this goal "it is crucial that the power sector is almost fully decarbonised by 2030". It has said that power sector emissions should be cut so that no more than 50 g of carbon dioxide are produced per kWh of electricity generated. Friends of the Earth's pathway meets this target.

## 6.2   Heat

Clearly the power sector is not the only, or indeed the main, sector.[81] In 2012, 40% of UK gas use was in the domestic sector; accordingly, our dependency on gas must also include reducing the use of gas in the domestic sector, or changing how this gas is sourced.

This should involve:

- *Improving home energy efficiency.* A recent report from the Association for the Conservation of Energy suggested that a home with an average level of energy efficiency could save £313 a year (or 25% or their heating bill) and an average fuel-poor home could save £650 a year through energy efficiency measures.[82] While some people are forced to choose between heating and eating due to poorly insulated homes, funding for energy efficiency schemes has reduced considerably since 2012. The only way to cut gas usage effectively, and deal with fuel poverty in the long-term, is through a large-scale, publicly funded energy-efficiency programme.
- *Development of renewable gas.* National Grid calculated in 2009 that renewable gas (produced mainly *via* anaerobic digestion or thermal gasification of biodegradable waste) could meet up to 50% of the UK's residential gas demand.[83] However, care needs to be taken about whether the feedstock could be better dealt with by other means, such as recycling, and the use of agricultural crops for anaerobic digestion needs to be discouraged.
- *Electrification of heat.* Early decarbonisation of power generation allows the use of electricity for the decarbonisation of other sectors, such as heating through the use of ground- and air-source heat pumps. National Grid has said that this is a "critical component of decarbonising heat".[84]

## References

1. Prime Minister's Office, *Local Councils to Receive Millions in Business Rates from Shale Gas Developments,* 2014; https://www.gov.uk/government/news/local-councils-to-receive-millions-in-business-rates-from-shale-gas-developments (last accessed 13/01/2014).
2. The Spectator, *Conservative Conference: Owen Patterson says Yes to Shale Gas and No to Wind Farm Subsidies,* 2012; http://blogs.spectator.co.uk/coffeehouse/2012/10/conservative-conference-owen-paterson-says-yes-to-shale-gas-and-no-to-wind-farms-subsidies/ (last accessed 01/12/2013).
3. European Commission, *Impact Assessment* accompanying the Communication from the Commission to the European Parliament, the Council, the European Economic and Social Committee and the Committee of the Regions on *Exploration and Production of Hydrocarbons (such as Shale Gas) using High Volume Hydraulic Fracturing in the EU*, 2014.
4. DECC, *The Myths and Realities of Shale Gas Exploration*, 2013; https://www.gov.uk/government/speeches/the-myths-and-realities-of-shale-gas-exploration (last accessed 14/01/2014).
5. AEA Technology for the European Commission, *Support to the Identification of Potential Risks for the Environment and Human Health arising from Hydrocarbons Operations involving Hydraulic Fracturing in Europe*, 2012.

6. UNEP, Gas Fracking: Can we Safely Squeeze the Rocks?, 2012.
7. Advertising Standards Authority, *ASA Adjudication on Cuadrilla Resources Ltd*, 2013; http://www.asa.org.uk/Rulings/Adjudications/2013/4/Cuadrilla-Resources-Ltd/SHP_ADJ_203806.aspx (last accessed 09/12/2013).
8. IPCC, *2013: Summary for Policymakers, in Climate Change 2013: The Physical Science Basis*, Contribution of Working Group I to the Fifth Assessment Report of the Intergovernmental Panel on Climate Change, ed. T. F. Stocker, D. Qin, G.-K. Plattner, M. Tignor, S. K. Allen, J. Boschung, A. Nauels, Y. Xia, V. Bex and P. M. Midgley, Cambridge University Press, Cambridge, UK, and New York, NY, USA.
9. Commission of the European Communities, *Limiting Global Climate Change to 2 degrees Celsius. The way ahead for 2020 and beyond*, 2007.
10. J. McCarthy, O. Canziani, N. Leary, D. Dokken and K. White, *Climate Change 2001: Impacts, Adaptation, and Vulnerability*, Contribution of Working Group II to the Third Assessment Report of the Intergovernmental Panel on Climate Change, Cambridge University Press, 2001.
11. J. B. Smith, S. H. Schneider, M. Oppenheimer, G. W. Yohe, W. Hare, M. D. Mastandrea, A. Patwardhan, I. Burton, J. Corfee-Morlot, C. H. D. Magadza, H.-M. Füssel, A. B. Pittock, A. Rahman, A. Suarez and J.-P. van Ypersele, Assessing dangerous climate change through an update of the Intergovernmental Panel on Climate Change (IPCC) "reasons for concern", *Proc. Natl. Acad. Sci. U. S. A.*, 2009, **106**(11), 4133–4137.
12. Carbon Tracker Initiative, *Unburnable Carbon – Are the World's Financial Markets carrying a Carbon Bubble?*, 2012; http://www.carbontracker.org/site/wp-content/uploads/2014/05/Unburnable-Carbon-Full-rev2-1.pdf (last accessed 06/06/2014).
13. The Guardian, *The Peak Oil Brigade is leading us into Bad Policy making on Energy*, 2011; http://www.theguardian.com/commentisfree/2011/oct/18/energy-price-volatility-policy-fossil-fuels (last accessed 09/12/2013).
14. David Hone, Climate Change Advisor for Shell, *The Carbon Bubble Reality Check*, 2013; http://blogs.shell.com/climatechange/2013/05/bubble/ (last accessed 10/12/2013).
15. DECC, Written Ministerial Statement by Edward Davey, *Review of UK Offshore Oil and Gas Recovery*, 2013; https://www.gov.uk/government/speeches/written-ministerial-statement-by-edward-davey-review-of-uk-offshore-oil-and-gas-recovery (last accessed 07/01/2014).
16. Friends of the Earth, *The UK, Shale Gas and Unburnable Carbon: Questions for the Government*, 2013.
17. M. Meinshausen, N. Meinshausen, W. Hare, S. Raper, K. Frieler, R. Knutti, D. Frame and M. Allen, Supplementary information – Greenhouse gas emission targets for limiting global warming to 2 °C, *Nature*, 2009, **458**, 1158–1162, doi: 10.1038/nature08017.
18. Using data from J. Olivier, G. Janssens-Marhout and J. Peters, *Trends in Global $CO_2$ Emissions 2012 Report*, PBL Netherlands Environmental Assessment Agency, The Hague, 2012. (The subsequent figures are very similar to those in Meinshausen et al.[17]).

19. Reserves and conversion data from BP, *Statistical review of World Energy 2013*, 2013.
20. International Energy Agency, *Golden Rules for a Golden Age of Gas*, 2012.
21. BBC, *Campaigners' Anger over Agency's Shale Gas Report*, 2012; http://www.bbc.co.uk/news/science-environment-18236535 (last accessed 12/01/2014).
22. United Nations, *United Nations Framework Convention on Climate Change*, 1992.
23. I. J. Andrews, *The Carboniferous Bowland Shale Gas Study: Geology and Resource Estimation*, British Geological Survey for Department of Energy & Climate Change, London, UK, 2013.
24. Committee on Climate Change, *The Need for a Carbon Intensity Target in the Power Sector*, 2012.
25. P. Stevens, *Why Shale Gas Won't Conquer Britain*, 2014; http://mobile.nytimes.com/2014/01/15/opinion/why-shale-gas-wont-conquer-britain.html (last accessed 17/01/2014).
26. J. Broderick, K. Anderson, R. Wood, P. Gilbert, M. Sharmina, A. Footitt, S. Glynn and F. Nicholls, *Shale Gas: An Updated Assessment of Environmental and Climate Change Impacts*, A report commissioned by The Co-operative and undertaken by researchers at the Tyndall Centre, University of Manchester, 2011.
27. EurActiv, *Shale Gas Strategy 'Not the Optimum Path': Fatih Birol*, 2012; http://www.euractiv.com/energy/fatih-birol-gas-definitely-optim-news-513043 (last accessed 01/12/2013).
28. J. Broderick and K. Anderson, *Has US Shale Gas Reduced $CO_2$ Emissions?*, Tyndall Centre, Manchester, 2012.
29. DECC, *Digest of UK Energy Statistics*, 2013.
30. Carbon Tracker Initiative, *Unburnable Carbon 2013: Wasted Capital and Stranded Assets*, 2013.
31. Chartered Institute of Water & Environmental Management, *Shale Gas and Water*, 2014.
32. Water UK, *Briefing Paper – Impacts of the Exploration for and Extraction of Shale Gas on Water and Waste Water Service Providers*, 2013.
33. C. G. Groat and T. W. Grimshaw, *Fact-Based Regulation for Environmental Protection in Shale Gas Development*, University of Texas at Austin, 2012.
34. Real Clear Energy, *Gasland II and Anti-Energy Extremists*, 2013; http://www.realclearenergy.org/articles/2013/06/19/gasland_ii_and_anti-energy_extremists.html (last accessed 15/01/2014).
35. US Environmental Protection Agency, *Draft Report – Investigation of Ground Water Contamination near Pavillion, Wyoming*, 2011.
36. ProPublica, *EPA's Abandoned Wyoming Fracking Study One Retreat of Many*, 2013; http://www.propublica.org/article/epas-abandoned-wyoming-fracking-study-one-retreat-of-many accessed 01/11/2013).
37. Bloomberg, *Drillers Silence Fracking Claims with Sealed Settlements*, 2013; http://www.bloomberg.com/news/2013-06-06/drillers-silence-fracking-claims-with-sealed-settlements.html (last accessed 01/11/2013).

38. EcoNews, *Industry Word Games Mislead Americans on Fracking*, 2013; http://ecowatch.com/2013/11/08/industry-mislead-americans-on-fracking/ (last accessed 10/11/2013).
39. S. G. Osborn, A. Vengosh, N. R. Warner and R. B. Jackson, Methane contamination of drinking water accompanying gas-well drilling and hydraulic fracturing, *Proc. Natl. Acad. Sci. U. S. A.*, 2011 ; published ahead of print May 9, 2011; doi:10.1073/pnas.1100682108.
40. USA Today, *4 States Confirm Water Pollution from Drilling*, 2014; http://www.usatoday.com/story/money/business/2014/01/05/some-states-confirm-water-pollution-from-drilling/4328859/ (last accessed 12/01/2014).
41. T. Colborn, C. Kwiatowski, K. Schultz and M. Bachran, Natural gas operations from a public health perspective, *Hum. Ecol. Risk Assess.: Int. J.*, 2011, **17**(5), 1039–1056.
42. C. D. Kassotis, D. E. Tillitt, J. W. Davis, A. Hormann and S. C. Nagel, Estrogen and androgen receptor activities of hydraulic fracturing chemicals and surface and ground water in a drilling-dense region", *Endocrinology*, **155**(3), 897–907.
43. Public Health England, *Review of the Potential Public Health Impacts of Exposures to Chemical and Radioactive Pollutants as a Result of Shale Gas Extraction – Draft for Comment*, 2013.
44. L. M. McKenzie, R. Guo, R. Z. Witter, D. A. Savitz, L. S. Newman and J. L. Adgate, Birth outcomes and maternal residential proximity to natural gas development in rural Colorado, *Environ. Health Perspect.*, 2014; doi: 10.1289/ehp.1306722.
45. L. M. McKenzie, R. Z. Witter, L. S. Newman and J. L. Aldgate, Human health risk assessment of air emissions from development of unconventional natural gas resources, *Sci. Total Environ.*, 2012; doi: 10.1016/j.scitotenv.2012.02.018.
46. Physicians, Scientists & Engineers for Healthy Energy, *Impediments to Public Health Research on Shale (Tight) Oil and Gas Development*, 2013.
47. M. Finkel, J. Hays and A. Law, The shale gas boom and the need for rational policy, *Am. J. Public Health*, 2013, **103**, 1161–1163.
48. *Daily Telegraph*, Cuadrilla admits drilling caused Blackpool earthquakes, 2011; http://www.telegraph.co.uk/finance/newsbysector/energy/8864669/Cuadrilla-admits-drilling-caused-Blackpool-earthquakes.html (last accessed 14/01/2014).
49. C. A. Green, P. Styles and B. J. Baptie, *Preese Hall Shale Gas Fracturing: Review and Recommendations for Induced Seismic Mitigation"*, 2012.
50. *The Times, Blackpool tremors reopen questions over fracking*, 2012; http://www.thetimes.co.uk/tto/business/industries/naturalresources/article3310081.ece (last accessed 01/10/2013).
51. *StateImpact, How Fracking Causes Earthquakes, But Not the One in Virginia*, 2011; http://stateimpact.npr.org/pennsylvania/2011/08/26/how-fracking-causes-earthquakes-but-not-the-one-in-virginia/ (last accessed 14/01/2014).

52. *Kent Online*, Kent outlined as key battleground for fracking at four sites in Tilmanstone, Guston Court Farm, Shepherdswell and Woodnesborough, 2013; http://www.kentonline.co.uk/east_kent_mercury/news/fracking-our-future-hope-or-8502/ (last accessed 14/01/2014).

53. National Audit Office, *Infrastructure Investment: The Impact on Consumer Bills*, 2013.

54. Committee on Climate Change, *Household Energy Bills – Impacts of Meeting Carbon Budgets*, 2011.

55. *Daily Telegraph*, *We cannot afford to miss out on shale gas*, 2013; http://www.telegraph.co.uk/news/politics/10236664/We-cannot-afford-to-miss-out-on-shale-gas.html (last accessed 14/01/2014).

56. DECC, *The Myths and Realities of Shale Gas Exploration*, 2013; https://www.gov.uk/government/speeches/the-myths-and-realities-of-shale-gas-exploration (last accessed 14/01/2014).

57. *The Independent*, 'Baseless economics': Lord Stern on David Cameron's claims that a UK fracking boom can bring down price of gas, 2013; http://www.independent.co.uk/news/uk/politics/baseless-economics-lord-stern-on-david-camerons-claims-that-a-uk-fracking-boom-can-bring-down-price-of-gas-8796758.html (last accessed 12/01/2104).

58. *The Guardian*, Lord Browne: fracking will not reduce UK gas prices, 2013; http://www.theguardian.com/environment/2013/nov/29/browne-fracking-not-reduce-uk-gas-prices-shale-energy-bills (last accessed 14/01/2014).

59. Chatham House, *The Shale Gas Revolution: Hype and Reality*, 2010.

60. International Energy Agency, *Golden Rules for a Golden Age of Gas*, 2012.

61. *New Scientist*, The UK's new dash for gas is a dangerous gamble, 2012; http://www.newscientist.com/article/dn22594-the-uks-new-dash-for-gas-is-a-dangerous-gamble.html (last accessed 12/01/2014).

62. *Bloomberg New Energy Finance*, UK shale gas no "Get Out Of Jail Free" card, 2013; http://about.bnef.com/press-releases/uk-shale-gas-no-get-out-of-jail-free-card/ (last accessed 15/12/2013).

63. Regeneris Consulting, *Economic Impact of Shale Gas Exploration & Production in Lancashire and the UK*, 2011.

64. Institute of Directors, *Getting Shale Gas Working*, 2013.

65. *Industry Week*, The great debate over shale gas employment figures, 2011; http://www.industryweek.com/public-policy/great-debate-over-shale-gas-employment-figures (last accessed 06/01/2014).

66. Regeneris research for Cuadrilla showing that the number of jobs created at around 1600 in Lancashire and 5600 in the UK for four years from 2016 to 2019, falling to under 200 from 2022 onwards.

67. *ENR New York, Hydrofracking offers Short-Term Boom, Long-Term Bust*, 2011; http://newyork.construction.com/opinions/viewpoint/2011/0307_HydrofrackingOffers.asp (last accessed 06/01/2014).

68. DECC, *Strategic Environmental Assessment for further Onshore Oil and Gas licensing – Environmental Report*, 2013.

69. *Bloomberg New Energy Finance*, UK shale gas no "Get Out Of Jail Free" card, 2013; http://about.bnef.com/press-releases/uk-shale-gas-no-get-out-of-jail-free-card/ (last accessed 15/12/2013).

70. European Commission Joint Research Centre, *Unconventional Gas: Potential Energy Market Impacts in the European Union*, 2012.

71. Professor Michael Bradshaw for Friends of the Earth, *Time to take our foot off the gas?*, 2012.

72. Australian Government Department of Climate Change and Energy Efficiency, *Coal Seam Gas Estimation and Reporting of Greenhouse Gas Emissions*, 2012.

73. National Toxics Network, *Toxic Chemicals in the Exploration and Production of Gas from Unconventional Sources*, 2013.

74. National Toxics Network, *Submission to New South Wales Inquiry into Coal Seam Gas*, 2011.

75. National Toxics Network, *Toxic Chemicals in the Exploration and Production of Gas from Unconventional Sources"*, 2013.

76. New South Wales Government, *Coal Seam Gas Exclusion Zones*, 2013; http://www.planning.nsw.gov.au/coal-seam-gas-exclusion-zones (last accessed 17/01/2014).

77. The Australian, *Underground Coal Gasification Plant near Kingaroy Shut Down after Cancer-causing Chemical found in bores*, 2010; http://www.theaustralian.com.au/news/underground-coal-gasification-plant-near-kingaroy-shut-down-after-cancer-causing-chemical-found-in-bores/story-e6frg6n6-1225892659672 (last accessed 17/01/2014).

78. Environment Agency, *Underground Coal Gasification*, 2010; http://webarchive.nationalarchives.gov.uk/20140328084622/http://www.environment-agency.gov.uk/static/documents/Business/UCG_factsheet_16_Aug10.pdf (last accessed 06/01/2014).

79. McLaren Environmental for the European Climate Foundation, *The Likely Implications for Climate Change from Development and Deployment of Underground Coal Gasification Technologies*, 2012.

80. Friends of the Earth, *A Plan for Clean British Energy*, 2012.

81. DECC, *UK Energy in Brief 2013*, 2013.

82. Association for the Conservation of Energy/Energy Bill Revolution, *Burning Cash Day*, 2014; http://www.energybillrevolution.org/wp-content/uploads/2014/02/ACE-and-EBR-briefing-2014-02-06-Burning-Cash-Day.pdf (last accessed 14/02/2014).

83. National Grid, *The Potential for Renewable Gas in the UK*, 2009.

84. National Grid, *Pathways for Decarbonising Heat*, 2012.

# Subject Index